Green Skills Research in South Africa

This book proposes transformative, realist methodology for skills research and planning through an analysis of case studies of the changing world of work, new learning pathways and educational system challenges.

Studies of the green economy and sustainability transitions are a growing field internationally, however there are few books that link this interest to the development of skills. This book draws on, and showcases, the experience and insights of researcher-practitioners who are at the cutting edge in this emerging field, internationally and in South Africa. The context for this book is South Africa, but application is worldwide. In many ways indicative of the global picture, South Africa is in the grip of economic and environmental imperatives, searching for safe and just transitions. The authors present a new, embedded transitioning systems model for studying skills for a sustainable, just future.

This book will be of great interest to students and scholars of sustainable development, ecological economics and skills planning.

Eureta Rosenberg is the Chair of Environment and Sustainability Education and Director of the Environmental Learning Research Centre at Rhodes University, Grahamstown, South Africa.

Presha Ramsarup is the Director at the Centre for Researching Education and Labour, University of Witwatersrand, Johannesburg, South Africa.

Heila Lotz-Sisitka is a DST/NRF Chair of Global Change and Social Learning Systems and a Distinguished Professor in the Environmental Learning Research Centre at Rhodes University, Grahamstown, South Africa.

Routledge Studies in Sustainability

Urban Social Sustainability
Theory, Practice and Policy
Edited by Ramin Keivani and M. Reza Shirazi

Holistic Sustainability through Craft-design Collaboration
Rebecca Reubens

Sustainable Development Teaching
Ethical and Political Challenges
Katrien Van Poeck, Leif Östman and Johan Öhman

Utopia in the Anthropocene
A Change Plan for a Sustainable and Equitable World
Michael Harvey

Sustainability Governance and Hierarchy
Philippe Hamman

Energy, Environmental and Economic Sustainability in East Asia
Policies and Institutional Reforms
Edited by Soocheol Lee, Hector Pollitt and Kiyoshi Fujikawa

Green Skills Research in South Africa
Models, Cases and Methods
Eureta Rosenberg, Presha Ramsarup and Heila Lotz-Sisitka

For more information on this series, please visit www.routledge.com/Routledge-Studies-in-Sustainability/book-series/RSSTY

Green Skills Research in South Africa

Models, Cases and Methods

Eureta Rosenberg, Presha Ramsarup and Heila Lotz-Sisitka

Routledge
Taylor & Francis Group
LONDON AND NEW YORK

earthscan
from Routledge

First published 2020
by Routledge
2 Park Square, Milton Park, Abingdon, Oxon OX14 4RN

and by Routledge
52 Vanderbilt Avenue, New York, NY 10017

Routledge is an imprint of the Taylor & Francis Group, an informa business

British Library Cataloguing-in-Publication Data
A catalogue record for this book is available from the British Library

Library of Congress Cataloging-in-Publication Data
Names: Rosenberg, Eureta, author. | Ramsarup, Presha, author. |
Lotz-Sisitka, Heila, 1965– author.
Title: Green skills research in South Africa : models, cases and methods /
Eureta Rosenberg, Presha Ramsarup and Heila Lotz-Sisitka.
Description: New York : Routledge, 2020. |
Includes bibliographical references and index.
Identifiers: LCCN 2019041140 (print) | LCCN 2019041141 (ebook) |
ISBN 9780367233402 (hardcover) | ISBN 9780429279362 (ebook)
Subjects: LCSH: Green technology–Study and teaching–South Africa. |
Occupational training–South Africa.
Classification: LCC HC79.E5 R655 2020 (print) |
LCC HC79.E5 (ebook) | DDC 338.927–dc23
LC record available at https://lccn.loc.gov/2019041140
LC ebook record available at https://lccn.loc.gov/2019041141

ISBN: 978-0-367-23340-2 (hbk)
ISBN: 978-0-429-27936-2 (ebk)

Typeset in Goudy
by Newgen Publishing UK

Contents

List of figures

List of tables

List of boxes

Authors' biographies

Lead authors

Eureta Rosenberg is Professor and Chair in Environment and Sustainability Education at Rhodes University, South Africa, where she also leads a research programme on Monitoring and Evaluation. She has a long-standing interest in research methodology for contexts of transformation and has contributed to green skills related research over two decades. This includes the National Green Skills Programme (2015–2019) and the development of the Biodiversity Human Capital Development Strategy for South Africa in 2010, with government and civil society organisations.

Presha Ramsarup is the Director for the Centre for Researching Education and Labour at the University of Witwatersrand, Johannesburg. She currently leads a research programme on Occupations and Skills and was joint co-ordinator of the National Green Skills Programme. She has served as a technical advisor to both provincial and national governments to support the development of environmental and sustainability education within South Africa. She was part of the core team to produce South Africa's first Environmental Sector Skills Plan and continues to work at supporting systemic engagement with green skills issues.

Heila Lotz-Sisitka is a Distinguished Professor in the Environmental Learning Research Centre at Rhodes University, South Africa. She holds a Tier 1 Department of Science and Technology/National Research Foundation Research Chair focusing on Transformative Social Learning and Green Skills Learning Pathways. She has been involved in various forms of green skills research over the years including curriculum research, research for environmental sector skills planning and a research programme for the South African Qualifications Authority on sustainable development work and learning, and green skills learning pathways.

Contributing authors

Najma Mohamed currently works as Policy Director at the Green Economy Coalition. With experience in research, policy and practice, she is a technical advisor and specialist on environment, development and climate change across economic sectors. Her recent experience has involved a focus on green economy, as a policy advisor to South Africa's Green Fund and as a specialist on just transitions and green jobs for the International Labour Organization. She has been active in the print and broadcast media and edited a volume on *Sustainability Transitions in South Africa* (2019), which assesses whether the country's transition pathway is both green and just.

Dr Tichaona Pesanayi (late) was a Southern African Development Community Researcher and Secretary General of the Environmental Education Association of Southern Africa.

Other contributors

The authors would like to thank all the researchers on whose work they have drawn in this book, and to acknowledge the South African Department of Environmental Affair's Green Fund (with the Development Bank of Southern Africa) for financial support to the National Green Skills Programme. Research in this book has also been supported in part by the South African Research Chairs Initiative of the Department of Science and Technology and National Research Foundation of South Africa (Grant No 98767), with specific reference to the work of the NRF SARChI Chair in Global Change and Social Learning Systems at Rhodes University. Much gratitude is also due to the team at Rhodes University, in particular Ms Ulisha van der Merwe, for helping to produce the book.

Prologue

Eureta Rosenberg, Heila Lotz-Sisitka and Presha Ramsarup

The destruction of ecosystems by economic activity that still fails to provide for the majority of planetary inhabitants forces us to re-think the work we do. New work and skills are needed to restore ecosystems and advance inclusive economies and society – a vision spread across 17 global Sustainable Development Goals (SDGs) (UN, 2015) – as are critical and transformative approaches to green skills research.

This book proposes a transformative, critical realist underlabouring for skills research methodology through an analysis of case studies of the changing world of work, new learning pathways, curriculum and learning as well as education system challenges. The context is South Africa, but the application can be worldwide. In many ways indicative of the global picture, South Africa is in the grip of economic and environmental imperatives that are mirrored elsewhere. Like others around the world, South Africans are searching for safe and just transitions to a more sustainable and socially just world order.

The case studies in the book emerge from three generations of green skills research undertaken within the context of a rapidly changing society, seeking to transform and transition out of the legacy of centuries of colonialism, racism, apartheid and extractivism. Intersecting with these social change imperatives are demands emerging from the ecological and climate crisis and the increasing realisation that there is a need for radically different, systemic alternatives. These are reflected in international, national and sectoral policies and intentions, including the SDGs and the South African National Development Plan (NPC, 2012) which points to the need for socially just transitioning to a low-carbon, job intensive and ecologically sustainable (i.e. 'green') economy.

Alternatives towards achieving the outcomes of the intended policy are slowly emerging, but as this occurs, the system reveals numerous challenges to change, contradictions and lock-ins in historic pathways. Some of these contradictions are conceptual and methodological in nature, inherited from skills systems and skills system research trajectories embedded in histories of industrial and colonial capital that largely failed to take into account the wider ecological system in which they are embedded (Moore, 2015). Coupled with more contemporary neo-liberal policies and trajectories, these limited trajectories shape skills systems thinking not only in South Africa, but worldwide. This legacy creates a complex context for conceptualising a fuller meaning of 'transformation' in skills system research, fraught with risks.

Most prominently, but not exclusively, the climate crisis characterised by a heating planet, irrevocably linked to human activity shaped by centuries of fossil and capital-based development, "unhinges all our certainties and places everything in jeopardy. It challenges our fixation with growth economies, 'catch up' development, and every conception of modern progress that has incited our imaginations" (Satgar, 2018, p. 1).

This book discusses the implications for skills planning and development that are induced by this contemporary condition. Paradoxically, and sadly, the contemporary condition most affects the excluded and marginalised members of society who are most vulnerable to economic and ecological failures. Ultimately, all humans, as well as the more-than-human inhabitants of the Planet, are affected by these issues. Skills development systems simply cannot ignore this.

Skills system and work and learning research trajectories

As can be seen from the research reported on and discussed across the pages of this book, our research on green skills and skills system development is located within an emergent, reflexive orientation and trajectory. It embraces an explicit re-thinking of work and learning research situated in the Anthropocene/Capitalocene[1] (Moore, 2015) and in the context of global commitments to sustainable development (most explicitly stated in, but not limited to, the SDGs).[2]

Our skills research, underlaboured by critical realism (cf. Chapter 13), is conceptualised within a Just Transitions framework which articulates a multi-levelled, emergent and relational view of transformation and change (cf. Chapter 1) that is not without significant contradictions and disruptive influences (cf. Chapter 14). This orientation to skills system research emerges from three generations of green skills research in which we have been involved over a 30-year period:

- **Generation 1:** Qualifications design and curriculum innovation research to address the absence of environmental qualifications and skills (early 1990s to the present)
- **Generation 2:** Work and learning research including learning pathways research into emerging green occupations and green work (mid-2000s to the present)
- **Generation 3:** Demand for green work and learning and green skills system building for the transformation of education, training and learning systems (2010 to the present).

In this book we avoid a technical use of 'skills'; instead we use the concept broadly, to denote a complex relationship between knowledge, skills, competences and values. Our work is situated within the wider skills development research and praxis landscape. It seeks to address a dualism between macro and micro-level studies which characterise the skills research landscape.

To develop the chapters in this book we have reviewed and worked within specific traditions of research within the skills research landscape (e.g. work and learning transitions research, demand and supply studies, learning needs assessments, occupational research, skills ecosystems research, curriculum and learning research, etc.). In all cases, we have found these traditions to be lacking adequate scope of engagement with the types of transformation issues required by global and national sustainable development challenges. This has produced a need for methodological development and innovation, which we share across the pages of this book.

What is green work and learning?

We use the concepts of 'green skills' and 'green work' judiciously in this book, realising that these are in some way metaphorical concepts, and open to multiple interpretations (cf. Chapters 2 and 12). To address this, we have developed a conceptual and methodological framework that has helped us to contextualise and deepen emancipatory perspectives on green skills. This framework embodies an explicit commitment to just transitions and its critical realist underlabouring (see Chapter 13) which addresses anthropocentric limitations in research and praxis, while maintaining a strong commitment to social, epistemic and environmental justice, and allows for situated and ethically constituted dialectical options for the emergence of transformative praxis. The framework allows for considering green work, the greening of work and green jobs within a critically constituted sustainable development framework (cf. Chapter 12).

Overview of the book

The book begins with an introduction to the transformation landscape that is shaping the emergence of green skills and the need for green skills research in South Africa and elsewhere (Chapter 1). This chapter makes the case for a transformative approach, with reference to the need for *radical* economic change that is regenerative towards achieving greater environmental sustainability and social justice at the same time. It provides conceptual framings for the change and change processes required, viewing the changes needed through a transitioning systems lens, with emphasis on Just Transitions (Swilling & Anneke, 2012; Satgar, 2018).

Following this is a chapter on the dynamic and changing policy landscape which shapes and influences the emergence of green skills and green skills research. This is a multi-dimensional policy landscape, and includes international policy trajectories such as the global Sustainable Development Goals (UN, 2015), multilateral environmental agreements, and global agreements such as the recent Paris Agreement to halt climate change (UNFCC, 2016). Influenced by these emergent policy trajectories and the issues themselves, national environmental policy is often characterised by 'newness' and rapid emergence, which has also been the case in South Africa. At the same time, and often in parallel, national contexts have also experienced significant education, training and post-school policy changes. This complex policy configuration often produces misalignments and contradictions, but also dynamism in the policy and practice system and landscape. This shapes the need for different types of green skills research (cf. Chapter 14), interesting empirical research questions and methodological directions, as illuminated across the chapters in this book.

Following the introductory section are three distinct yet related sections. The first section focuses on the changing nature of work and skills research suitable for investigating associated skills needs or demand. Each chapter illustrates a different feature of the methodological frameworks for interpreting and understanding the changing nature of work, which is an essential step in proactive skills planning. Here we illustrate methodologies such as the use of a laminated system for the identification and analysis of green work and associated skills needs (cf. Chapters 2, 3 and 4), hotspotting analysis for identifying where leverage points for prioritising green skills development exist (cf. Chapter 3), the extension and re-orientation of commodity value chains (cf. Chapters 2, 3 and 4) and empirical mapping of occupations and their articulation(s) at the micro level (cf. Chapter 4) to make such skills visible in systems of planning. The chapters also illustrate the diversity of contexts where green skills research is needed such as mining, agriculture and the chemical industry (amongst others).

The second section of the book focuses on new learning pathways into green work. This section shows that many of the learning pathways into green occupations are new, and in some contexts relevant green occupations are either absent, or not well articulated, affecting the emergence of learning pathways into green work. In this section, we consider some of the methodological approaches to understanding the complexities of transitioning from higher education into green occupations which typify some learning pathways into green work (cf. Chapter 7) and some of the systemic constraints and enablements that affect pathways into specialist environmental occupations (cf. Chapter 6). We also consider the potential of social ecosystemic approaches to green skills research and planning (cf. Chapter 8). Methodologically, we draw on capabilities theory, ecosystemic approaches to transitioning research and career stories coupled with critical realist institutional system analysis for understanding learning pathways into specialist occupations. Contextually we cover both high skills (e.g. environmental scientists) as well as elementary occupations in local municipalities and public works programmes.

The third section of the book addresses questions related to education and training and skills system building. Here we consider the methodological issues associated with framing and conducting a national learning needs assessment (cf. Chapter 9), and review emerging implications of green skills development for providers and curriculum development (cf. Chapter 10). Methodological approaches inform contextually framed qualifications and learning programme design, reflexive curriculum review and the mediation of new knowledge. The section draws attention to the need for approaches to curriculum and learning theory innovation to respond to the inter-disciplinary, interrelated and often contextually emergent nature of sustainability concerns and knowledge.

Capping the book are three chapters that underlabour the rest of the book. They are theoretical and address wider system building concerns such as the need for critical engagement with the dominant discourse of occupations in skills research and planning (cf. Chapter 12), methodology development (cf. Chapters 13 and 14) and policy implications (cf. Chapter 14). Chapter 13 offers a synthesis of green skills research methodology focusing on the critical realist philosophical underlabouring we have worked with to shape transformative green skills research, as well as skills system theory and methods (cf. Chapter 13). The last chapter synthesises types of green skills research, and points to policy and system development implications and the way in which a green skills focus reconstitutes the notion of work (cf. Chapter 14). These three chapters reiterate the need to think about skills research, planning and development within an open system in which meaning(s) are changing rapidly and fundamentally as society comes to grips with both the risk and limitations of historically dominant forms of development thinking and praxis.

Overall, the book offers methodological guidance for researchers interested in re-thinking, reorienting and transforming the skills, education and training landscape within the global drive for sustainable development, ushered in by the Rio Earth Summit in 2002 and more firmly (if still inadequately) articulated in the 2015 Sustainable Development Goals. In writing this book, we recognise that the directions for transformation outlined are also to be found in centuries-old First Nations' people's relational commitments to more benign human-nature relations. Such commitments are also found in contemporary transformative sustainability responses that take seriously the interrelatedness of humans and the ecosystems of which they are part, and the implications of this currently fractured relation for current and future generations. These perspectives inform our foundational methodological commitment to relational thinking, generativity and transformative education, training and skills development.

Notes

1 The Anthropocene is a concept coined by Earth System Scientists to signal human impacts are fundamentally changing the geological structure of the Earth Systems. Critiquing this framing from a political economy perspective, Moore (2015) suggested we need to historicise engagement with the concept of the Anthropocene by pointing to the pernicious impacts of capitalism on the Earth system, and in doing so, he offered the concept of the 'Capitalocene' to draw attention to the confluence of political economy and political ecology (cf. also Chapter 14).

2 UN, 2015, Resolution 70/1 of the United Nations General Assembly. These global goals adopted by all UN states, form part of the broader 2030 Agenda for Sustainable Development. Addressing social, economic and ecological concerns, they challenge the historic assumption that environmental care is necessarily an opposition to development needs (UN, 2015).

References

Moore, J.W. (2015). *Capitalism in the web of life: Ecology and the accumulation of capital.* London: Verso.

National Planning Commission (NPC). (2012). *The National Development Plan 2030.* Pretoria: Office of the Presidency.

Satgar, V. (2018). Chapter 1: The climate crisis and systemic alternatives. In Satgar, V. (Ed.). *The Climate Crisis: The South African and global democratic eco-socialist alternatives.* Johannesburg: Wits University Press.

Swilling, M., & Annecke, E. (2012). *Just transitions: Explorations of sustainability in an unfair world.* United Nations University Press.

United Nations (UN). (2015). Sustainable development goals. Agenda 2030. Retrieved from www.un.org/sustainabledevelopment/sustainable-development-goals

United Nations Framework Convention on Climate Change (UNFCC). (2016). Decisions adopted by the Conference of the Parties on Climate Change; Paris Agreement, December 2015. Retrieved from https://unfccc.int/resource/docs/2015/cop21/eng/10a01.pdf

1 Skills for just transitions to sustainability

An orientation

Eureta Rosenberg and Presha Ramsarup

In 2011, following the Seventeenth Conference of the Parties convened by the United Nations Framework Convention on Climate Change, held in Durban, South Africa, civil society organisations including the National Union of Metalworkers of South Africa (NUMSA) campaigned for the creation of *one million climate jobs*. These jobs, the Climate Coalition argued, would be based on renewable energy, sustainable transport, water saving and other measures on a new, low-carbon development path (AIDC, 2011). A clean, green and labour intensive development path was also the focus of a Green Economy Accord forged in November 2011 between government, business and labour (EDD, 2011).

In March 2018, however, in an apparent about-turn, NUMSA participated in a High Court interdict application to force the Minister of Energy to stop renewable energy projects. The reasons given included the need to save coal truck drivers' jobs (News24).

Why the change?

The recognition that environmental issues have a direct impact on both economic growth and social well-being makes it clear that unless we act fast and decisively environmental issues will exacerbate inequality, poverty and unemployment. It is no longer a choice between jobs and environmental health. It has to be both or we face the spectre of neither (Ward, 2018).

A transition towards a greener economy cannot be without challenges. Industries like coal mining and fossil based power generation will be adversely affected (Stern, 2018). New industries must be developed in a socially inclusive manner and workers will need re-skilling for different jobs. A new development path requires concerted interventions including aligned skills development. Despite the 2011 Green Economy Accord, planning for sustainability and associated skills has not yet taken place in South Africa on a significant scale. Failure to develop a new development path is critical; in the first quarter of 2019 South Africa's economy was declining by minus 3%, and mining and manufacturing were the industries most affected.

These factors are also the motivation behind this book, which makes a case not only for the work needed to put the world on an environmentally sustainable path, but for the skills needed to do this work. It argues specifically that the research to inform the development of skills for sustainable futures must be *transformative* in orientation. If skills research simply follows the economic development logic of the past, it will fail to remove the deadlock of *jobs versus environment*. Some would argue (e.g. Ashley, 2018) that union responses are not only a reflection of a lack of planning for a sustainability transition, but a failure to understand quite how radical a change is needed to make this transition; at stake is the very nature of the economic thinking on which so much of development planning, even some forms of sustainable development planning, is based. The narrow, extractivist and exploitative development path that served a minority of South Africans well during the apartheid years is failing to meet the needs of the majority, and the African National Congress and its majority rule government is yet to find an alternative that provides economic opportunities for all citizens.

This introductory chapter will argue that a transformative approach is necessary, involving *radical* economic change towards environmental sustainability and social justice – conceived of as two sides of the same coin, rather than opposing forces. It provides conceptual framings for the change processes required that are to be found in a relational philosophy and complex systems thinking applied to a broader vision of the economy, including a framing of environmental sustainability as a social justice issue, and viewing the changes needed through a transitioning systems lens. This forms the basis for a transformative approach to skills research, which will be demonstrated in the rest of the book.

Environmental sustainability matters

With a growing ability to use 'big data', earth scientists like Rockström et al. (2009) have illustrated that humanity's activities are exceeding several of the boundaries that make our planet habitable (see Figure 1.1). The gases produced by various industries are tipping the Earth into a new climate era to which we cannot adapt fast enough. We are polluting the air we breathe, the water we drink and the soils and oceans from which we feed. Those with the fewest choices and the least ability to change the course of events, the most economically marginalised humans (Stern, 2007) and the *more-than-human* species (Weston, 1994), are most at risk. There are therefore physical *and ethical* reasons for seeking environmental sustainability for our home planet. There are also economic reasons for seeking environmental sustainability as national and regional economies are deeply threatened by climate change, with the cities of Africa being among the most vulnerable.

Southern Africa is already severely affected by extreme climate events, and the region is likely to experience increasing and more severe droughts, floods, famines and disease burdens (IPCC, 2014). These predictions of risk, like Rockström et al.'s calculations of the planetary boundaries, are abstract. But the people of the region are already experiencing environmental issues. South Africa's neighbours, Mozambique and Zimbabwe, are periodically affected by tropical cyclones, but in 2019 two successive cyclones uncharacteristically hit the south-eastern coast of the continent, destroying crops, infrastructure and thousands of lives in one season. Every year, air pollution (caused inter alia by heavy industries) kills 20 000 South Africans, costing the economy nearly R300 million (World Bank, 2016). Biodiversity – comprised of thousands of creatures and their habitats – is destroyed in the quest for more mines,

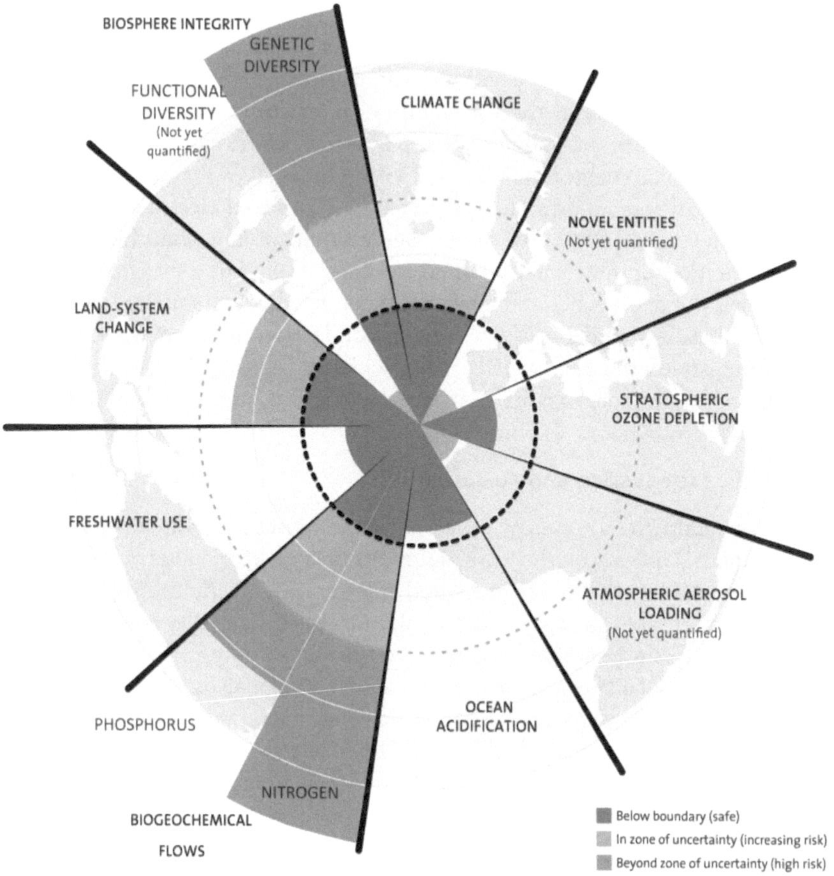

Figure 1.1 Planetary boundaries depicted
Source: J. Lokrantz/Azote based on Steffen et al. 2015

plantations, farmland and real estate. A loss of biodiversity has ethical and spiritual consequences, and also puts food production, tourism, water provisioning and other benefits in the economic 'commons' at risk. And despite all the extractive development, still we are unable to eradicate the human hardships caused by poverty. Many South African families do not have access to piped water or sanitation and live surrounded by untreated waste. While the country boasts world class shopping destinations frequented by its own millionaires, at the same time, an astonishing 38% of South Africans are unemployed (StatsSA, 2019, expanded unemployment definition). The country has the highest inequality metrics globally (Anand, Kothari & Kumar, 2016), but the situation is, sadly, a microcosm of the world at large.

In the face of dire socio-economic issues, environmental concerns have been deemed less important, and juxtaposed as being in competition with poverty eradication and employment creation. The shift required and evidenced by the One Million Climate Jobs Campaign and South Africa's Green Economy Accord, the National Development Plan as well as numerous national policies and international conventions is to see social development and environmental issues as intertwined, with roots in the socio-cultural practices of economics. The coal truck drivers' jobs are on the line because power stations are closing; it is becoming cheaper to produce energy from wind and solar sources (WWF, 2014), as accessible high grade coal sources dwindle, and water for coal-based electricity generation runs out (cf. Chapter 3). At the same time, it is difficult in a democracy to justify the exploitation of mineworkers whose toil earns them but a meagre living. As the demand for decent jobs and wages increases, mechanised mining becomes an attractive business model – as does energy production from renewable sources. This demonstrates the intertwined roots of social and ecological issues. In order to address such issues, solutions are being sought in new economic framings.

Economics for justice and sustainability

How do we address environment-development issues, if their roots are intertwined in economics? Drawing on the work of some of the best economists of the twenty-first century, Raworth (2017) presented a cogent argument that by changing some fundamental economic principles and practices, humanity could shift into an 'operating space' that is *both* ecologically safe and socially just (see Figure 1.2).

One example of addressing environmental and development issues concurrently involves the reduction of food waste. The volumes of food wasted during and after production are astronomical (WWF, 2017). By reducing this waste, more food can become more accessible to more people, thus reducing hunger and malnutrition *as well as* the amount of water, energy and wildlife habitat needed to produce 'food for all, forever'. The economic practices that make such waste viable and even desirable would need to change. Since economic practices are social constructions rather than physical inevitabilities, this should be possible. If we changed economic principles and practices that support food waste, we would address ecological *and* social imperatives.

Can this be done or is it a pipedream? After reviewing the work of ecological and environmental economists, Raworth (2017) distilled seven principles through which societies can address both the exceeding of planetary boundaries (as outlined by Rockström et al., 2009) and the failure to meet basic social needs, like decent work, access to water and energy, equitably. These principles include:

Systemic thinking and practices. We need to recognise that our development actions and economic activity affect our natural resources in one way or another, and that these resources are also a form of capital (while also being more than economic resources, including sources of spiritual connection). Applying systems thinking, we recognise that shared natural resources are an integral part of the economic equation, and that investing in the protection of natural resources adds to our collective capital. In 1993, following the first United Nations Summit on Environment and Development (also known as the Earth Summit), and drawing on the early Green Economy work of Ekins (1992), O'Donoghue depicted the interconnected nature of the systems affecting our daily lives (see Figure 1.3).

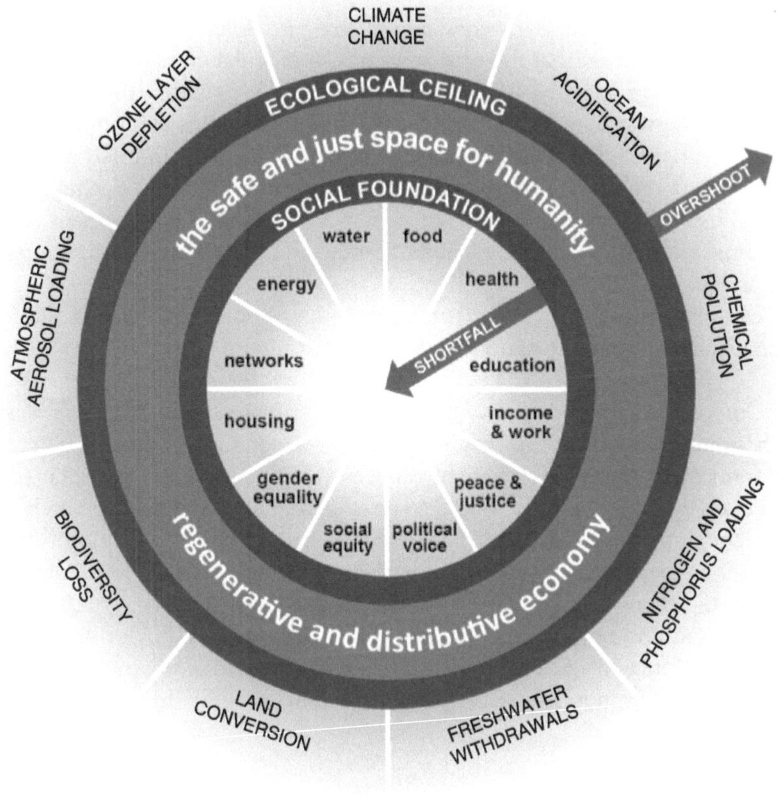

Figure 1.2 The possibility of operating in a safe and just space
Source: Raworth, 2017

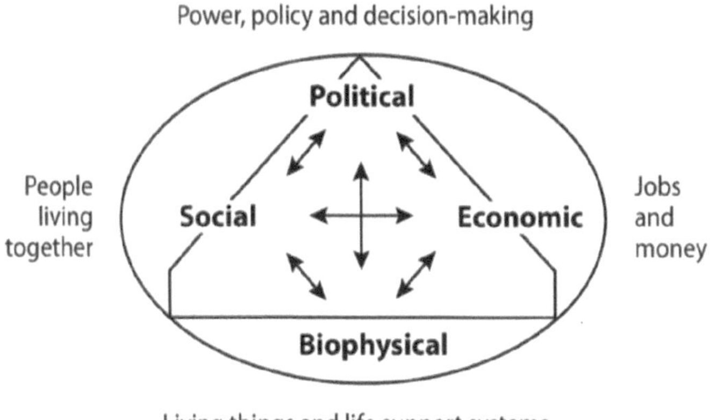

Figure 1.3 A systems perspective on environment
Source: O'Donoghue, 1993

'Seeing the big picture' is a second and related principle of systems thinking. It involves a recognition that the economy consists of and requires markets but also households (and associated food security, for example), regulatory functions (such as water conservation measures) and the commons (including air, water, land and human creativity), and that these need to function optimally in relation to each other for economies to thrive sustainably, to everyone's benefit. 'Seeing the big picture' has implications for how we view work worthy of employment status, including voluntary work, for example, to protect common natural resources. It even influences accounting.

Resource economist Blignaut (pers. comm., 2018) noted the value of a shift to recognising that land restoration is a form of investment:

> One option (to fund environmental protection work) would be to redefine restoration not as a cost item limited to the expenditure sheet of a company or organisation, that is invariably subject to cost reduction and cost cuttings, but as an investment that builds an asset on a balance sheet. Fortunately, the accountancy profession has now recognised that, and the definition of an asset has been revised recently, to be defined in a way that allows natural capital to be considered an asset. The moment that one follows a balance sheet approach and not an expenditure account approach, ecological restoration … is no longer about cost cutting, but about maximising the investment and the returns from it. It is a completely different space for environmental and economic practices.
>
> (See also Blignaut, 2017)

Regeneration: Instead of developing economies and creating wealth solely by uncontrollably extracting resources and causing degradation, we can aim to design wealth creation processes that are by nature frugal and restorative. Wahl (2016) has described regenerative cultures and argued that we stand at the dawn of a regenerative century. While this is a poetic vision, it can indeed be practically observed in examples of circular or cyclical economic practices which have been gaining traction internationally (www.ellenmacarthurfoundation.org) and in South Africa where an Industrial Waste Exchange (IWEX) programme forms part of an industrial symbiosis initiative of the Western Cape provincial government (O'Carroll et al., 2014). A regenerative business or industry goes beyond recycling; its processes are designed to reduce waste from the choice of materials and processes onwards, and their interconnected nature means that outputs from one process become inputs into another.

While economic growth may be needed in some periods and contexts, a further principle of living in a safe, sustainable space is to *be agnostic about economic growth per se*. Growth in the renewable energy sector might be accompanied by contraction in the fossil fuel sector, and show up as no-growth on a nation's economic balance sheet, while still benefiting people and planet.

The next two principles are strongly aligned.

Change the general goal of economics to achieving an ecologically safe and socially just space, rather than simply to target growth in Gross Domestic Product (GDP). Many critics have documented the perverse impacts a mindless pursuit of a higher GDP can have on countries and communities. Among these critics are Fioramonti (2013, 2017) and other proponents of well-being economies, who demonstrate that a growth in GDP sometimes worsens the lot of the poorest people, and deepens inequality. The goal of achieving human well-being re-directs the focus of economic planning and practice to improve, in the first instance, conditions for the most marginalised and vulnerable members of society.

Complementing these principles is the call to intentionally design global, national and regional economies for the *distribution of wealth*, rather than to assume that growth will result in a trickle-down effect that will eventually benefit the poor. Examples of such measures include wealth taxes and basic income grants (Standing, 2010). One of the strongest critiques of capitalism as our mainstream economic theory is that it benefits the few who are already well endowed, and exploits the rest to varying degrees, including the environment. This critique is strongly made by 'radical Marxists' like Ashley, Cock and Satgar (in Satgar, 2018). These authors also argue that some versions of 'green economy' are merely efforts to maintain the status quo (cf. Chapter 12), and are in fact dangerous for people and the planet, as they simply commodify more natural resources for profiteering by the few, with little attention to planetary boundaries, the 'safe and socially just operating space', systems thinking, circularity, regeneration and the other principles noted here.

Finally, Raworth also suggested that radically transformed economic principles and practices require us to foreground and *cultivate humanity's tendencies to cooperate with*, nurture and support others, rather than those tendencies to act based on self-interest and rational reasoning alone, what economic theory calls *Homo economicus*, the ultimate rational selfish human being, and on which mainstream economic theory is based. Research has shown that students on Economics courses that emphasise competition become more inclined to act like *Homo economicus* (Gilovich & Regan, 1993). However, humans are *both* cooperative and competitive, *both* sharing and selfish (Schwarz, 1994).

Humanity surely does have the creativity to design different social, cultural and economic systems over time, suitable for different contexts. Nonetheless, the tenets of mainstream economics are deeply entrenched and shifting them requires considerable effort. The shifts required are at the level of what systems theorist Donnella Meadows (1999) called the paradigmatic level, where worldviews are shaped. However, they also feature at the level of measurements and indicators, like GDP; and at the micro-level of everyday experience; that is, they are multi-levelled, system wide changes. This implies that if we are to study the work and skills needed for and within such transformations, we also need to pay careful attention to the paradigmatic assumptions, approaches, measurements and outcomes that we apply to and via our research methodology. This is a key theme in this book.

The way forward is not in any way self-evident; the cogent critiques and proposals by Moore (2015), Satgar (2018), Blignaut (2017), Standing (2010), Raworth (2017), Fioromonti (2017) and others provide only pointers, not blueprints. The pathway(s) to sustainability are likely to require extensive experimentation, adaptation and expansive social learning (cf. Chapters 8 and 11). Nor is this process of learning-by-doing likely to be smooth, universally supported and uncontested, and there will be resistance even from those whose improved well-being is an integral concern, as the vignette at the start of this chapter demonstrates.

In addition to knowing that environmental sustainability matters alongside social justice as an inseparably connected concern, and that radically different economic principles and practices are necessary for solving these socio-ecological concerns, we also need to think about *how* such radical changes might come about, or fail to come about.

For this we turn to the next part of this chapter, which introduces our framing of socially just sustainability transitions in complex systems.

Framing just transitions

Academics, governments and society broadly are formulating responses to the twin contemporary challenges of human development and environmental degradation that are causing humanity to overshoot planetary boundaries, while still failing to meet everyone's basic needs. Many, including the authors quoted earlier, have mapped the kinds of changes that may be needed. But if we are to use this information to understand the new forms of work and new skills that are needed, we also need a framing for *how* such changes might happen.

Sustainability transitions theory provides frameworks that enable insight into how change occurs in organisations, industries, sectors and other systems as they move to greater levels of sustainability over time, in response to innovations and other drivers. We have found value in the works of Geels (2010), Rotmans and Loorbach (2009) and Bhaskar and Danermark (2006) which we briefly introduce here as part of the framing of the cases and analysis in this book.

In general terms, a transition is a process of change from one state to another. Rotmans, Kemp and van Asselt (2001) defined 'transitions' as "transformation processes in which society changes in a fundamental way over a generation or more. [It is a] gradual, continuous process [of] change where the structural character of a society (or a complex sub-system of society) transforms" (pp.15–16).

The cases in this book are inherently about understanding the co-evolution of the sub-systems of the planet, industry and economics, workplaces and education towards a transition pathway to greater sustainability. While complex systems thinking necessitates that we view transitions as emergent processes, that are not entirely or even greatly predictable, planning is nonetheless needed which, in a social setting, can be guided by best available understandings of social and organisational change, and in education, by best available understandings of education and learning i.e. complexity thinking does not just licence the emergence of 'anything that can happen'; it allows for reasonable deliberation in the social sphere, with recognition that such knowledge is always fallible, and that ontological conditions shape such knowledge.

The works of Geels (2010) and Rothmans and Kemp (2003) describe a socio-technical, multi-level perspective (MLP) on sustainability transitions. This analytical framework is useful for outlining the layered nature of transitioning systems, where innovation and other drivers for change might occur, and where resistance can be expected (Geels & Kemp, 2012).

The first layer to consider is the middle level of three, described as *regimes*, as this is where change is often resisted. It includes institutional rules and regulations, e.g. related to industrial and economic practices and development models, and their ascendency over social equity and ecological considerations. It is here that one witnesses the lock-ins such as South Africa's minerals-energy complex (Fakir, 2017).

Comprising a second layer in the transitioning systems model, *landscapes* are broader than regimes, and include the drivers which may put pressure on regimes to shift. Examples of such drivers are ecological scarcities and risks (such as climate change, and dwindling mineral or water resources) that put pressure on particular development models (cf. SACMRM, 2013). Growing social discontent in the form of civil society protests over lack of access to water or jobs could be another landscape level driver encouraging government, business and civil society to invest in a green economy.

At the third level in the MLP framing are *niches*, that is, pockets of innovation and new practices that can act as sites of transformation, with the potential to stimulate and guide regime shifts (cf. Chapters 8, 11 and 14).

These transitions require new knowledge and skills both within existing sectors and occupations and in emerging sectors and occupations. Hence this book resonates with calls for an interdisciplinary perspective on skills, like that of Green (2013) who argues skills have three key features: they are productive, expandable and social. His argument positions skill as central to human, social and economic progress, and central to a discussion of social and economic action but fails to take the ecological condition in which society and economy are embedded into account.

This book draws on and expands (cf. Chapter 14) this broader framing of the notion of skill and contributes to methodology for investigating skills for a just transition. It argues that in order to understand what these 'skills' transitions may look like, we need to learn how to shift from minimal just transitions to transformative just transitions, from green resilience to green revolutions (cf. Chapter 12) and from instrumental business cases for profit maximisation to transformative business models for sustainable value creation (Ward, 2018). The chapters in this book provide specific insights into the principles of these transitions, how associated processes and practices may take place and what kinds of occupations and skills will be needed in particular sectors and contexts.

Typical socio-technical transition discourses do not expand extensively on the nature of the learning, interactions and transformations that contribute to the development of new practices, innovations and their wider uptake, or on the social learning that is needed to deal with the unintended consequences or partial failures of ongoing social change. For this the work of researchers at DRIFT including Rotmans and Loorbach (https://drift.eur.nl/research/) is useful. Building on the analytic frameworks of the multi-level perspective, the DRIFT group (ibid.) focuses on opportunities for transformative change starting with the notions of complexity, uncertainty and unpredictability, and combines systems thinking with relational approaches to place contemporary challenges in their historical context and in relation to potential futures. They also foreground issues of governance, power and agency in the roles of diverse actors.

To further the research and analysis of transitioning systems and associated work and skills, the studies in this book also applied the philosophical-ontological underlabourer of critical realism (Bhaskar, 1993; cf. Chapter 11). Underlabouring the transitioning work with critical realism enabled philosophical engagement to help us better understand the complex nature of environmental learning pathways as several ontologically different and irreducible levels of reality (Ramsarup, 2017; Lotz-Sisitka & Ramsarup, 2017). The critical realist approach prompts researchers to aim for a rich variety of perspectives on the same object and an examination of different dimensions of phenomena. This results in what Bhaskar called 'laminated explanations'. The studies in this book exemplify a need to surface and understand complexities as a laminated system for transitioning individuals, work, occupations and society towards greater sustainability (see Figure 1.4).

Conceiving of complex sustainability transitions as a laminated totality helps researchers contend with reductionism, as no one ontological level should be privileged in research, for example, the individual is not more important than structure or, green skills transition is not only about the development of new courses or qualifications. Nunez (2014) suggested that there is no *a priori* way in which reality produces complexity; thus the building of a laminated system does not follow a particular ordered approach; rather its formation is in direct relation to the explanatory concerns and the reality that determine them. Chapter 6 shares the laminated framework represented in Figure 1.4, reconceptualised for studying pathways of environmental scientists.

Drawing on critical realism not only helps to deepen our analytic research and give us a fuller perspective, it also helps us to move from description and analysis to a critical, generative and transformative approach.

Adopting a transitioning systems perspective without a critical realist lens could have resulted in simply describing *what is* happening and seeing this as adequate. Adopting a *critical* realist and therefore dialectical vantage point in our research (cf. Chapter 13) has helped to identify what is happening and how, but *also* to point more clearly to the absences i.e. what is *not (yet) happening* and how it might be possible to more proactively develop the system. It thus moves from a purely descriptive research process to critical and generative research processes for a *system development vantage point.*

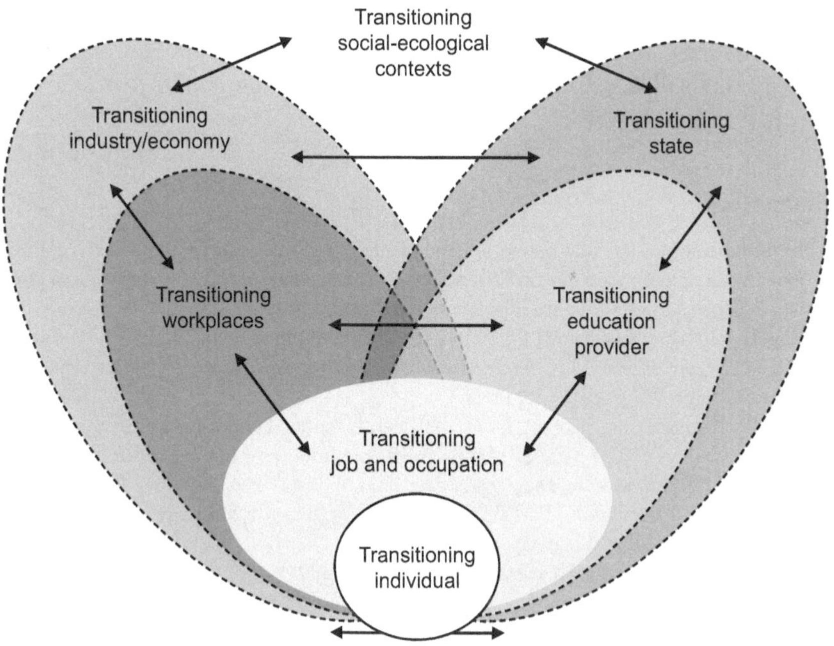

Figure 1.4 Overlapping transitioning systems for developing green skills
Source: Ramsarup et al., 2018

We devised the overlapping transitioning systems diagram (Figure 1.4) to help us approach and design research into the skills and capabilities needed to support the transition to a greener economy (Ramsarup, 2017; Ramsarup et al., 2018). Based on the transitioning systems literature we knew that we needed to understand the required changes at multiple analytic levels, which are constellationally related to all the other levels in the transitioning system. This relational, multi-level (laminar) perspective of transitioning acts as a conceptual framing for studying the 'greening' of work as the interplay at different analytical levels, as follows:

- The level of the transitioning workers – whose job, occupational tasks, knowledge and competences are changing
- The level of the transitioning nature of work, systems, jobs and workplace practices, where we may find evidence of the lock-ins and opportunities for transformative praxis
- The level of the transitioning sector, facing new production and consumption challenges, changing value positions, operating norms and practices, to explore the lock-ins
- The transitioning post-schooling system of education and training
- A transitioning society oriented towards sustainable development and a green economy, which applies landscape level pressures (Ramsarup, 2017).

The levels were conceived as transitioning in acknowledgement of the rapidly changing nature of the environmental sector and of the nature of work in the sector.

This framing, also introduced in Ramsarup et al. (2018), has been developed through, and will be used to introduce, the empirical case studies that follow in the chapters in this book.

References

Alternative Information and Development Centre (AIDC). (2011). *One million climate jobs: A just transition to a low carbon economy to combat unemployment and climate change*. Cape Town: Million Climate Jobs Campaign South Africa.

Ashley, B. (2018). Chapter 13: Climate jobs at two minutes to midnight. In V. Satgar (Ed.), *The climate crisis: The South African and global democratic eco-socialist alternatives*. Johannesburg: Wits University Press.

Anand, R., Kothari, S., & Kumar, N. (2016). South Africa: Labour market dynamics and inequality. IMF Working Paper 16/137. Retrieved from www.imf.org/external/pubs

Bhaskar, R. (1993). *Dialectic: The pulse of freedom*. London: Verso.

Bhaskar, R., & Danermark, B. (2006). Metatheory, interdisciplinarity and disability research: A critical realist perspective. *Scandinavian Journal of Disability Research*, 8(4), 278–297. https://doi.org/10.1080/15017410600914329

Blignaut, J.N. (2017). Chapter 37: The economics of restoration and the restoration of economics. In S.K. Allison & S.D. Murphy (Eds.), *The handbook of ecological and environmental restoration*, pp. 537–550. London and New York: T&F Routledge.

Cock, J. (2013). 'Green economy': A sustainable development path of 'wolf in sheep's clothing'? Paper presented at the Conference on Land Divided: Land and South African Society in 2013 in Comparative Perspective, 24–27 March, University of Cape Town, Cape Town.

Cock, J. (2018). Chapter 10: The climate crisis and a 'just transition' in South Africa: An eco-feminist-socialist perspective. In V. Satgar (Ed.), *The climate crisis: The South African and global democratic eco-socialist alternatives*. Johannesburg: Wits University Press.

Economic Development Department. (EDD) (2011). *New Growth Path: Accord 4: Green Economy Accord*. Pretoria: EDD.

Ekins, P. (1992). *A new world order: Grassroots movements for global change*. London; New York: Routledge.

Fakir, S. (2017). *Transition realism: The implications of rent-seeking to achieving South Africa's low-carbon technology ambitions*. Cape Town: WWF-SA.

Fioramonti, L. (2013). *Gross domestic problem. The politics behind the world's most powerful number*. London: Zed Books.

Fioramonti, L. (2017). *Wellbeing economy. Success in a world without growth*. London: Pan Macmillan.

Geels, F.W. (2010). Ontologies, socio-technical transitions (to sustainability), and the multi-level perspective. *Research Policy*, 39(4), 495–510.

Geels, F.W., & Kemp, R. (2012). The multi-level perspective as a new perspective for studying socio-technical transitions. Automobility in transition. In F.W. Geels, R. Kemp, G. Dudley, & G. Lyons (Eds.). *A socio-technical analysis of sustainable transport*, pp. 49–79. London: Routledge.

Gilovich, F.R., & Regan, T. (1993). Does studying economics inhibit cooperation? *Journal of Economic Perspectives*, 7(2), 159–171.

Green, F. (2013). *Skills and skilled work: An economic and social analysis*. London: Oxford University Press.

Intergovernmental Panel on Climate Change (IPCC). (2014). Climate Change 2014: *Synthesis Report*. *Contribution of Working Groups I, II and III to the Fifth Assessment Report of the Intergovernmental Panel on Climate Change*. Geneva: IPCC.

Lotz-Sisitka, H.B., & Ramsarup. P. (2017). Learning pathways and articulation: Early conceptual explorations and implications for research design(s). *SAQA Bulletin: Learning pathways, the National Qualifications Framework (NQF), and lifelong learning in South Africa, 17*(1): 31–48.

Moore, J. (2015). *Capitalism in the web of life: Ecology and the accumulation of capital*. London: Verso.

Nunez, I. (2014). *Critical realist activity theory: An engagement with critical realism and cultural-historical activity theory*. London: Routledge.

O'Carroll, S., Wallace, J., Pineo, C., Basson, L., Woodcock, J., Daniel, R., & Mouton, C. (2014). *The Western Cape Industrial Symbiosis Programme (WISP): An innovative approach to resource efficiency and waste minimisation for South African businesses*. Proceedings of the 20th WasteCon Conference, 6–10 October 2014. Institute of Waste Management, Cape Town.

O'Donoghue, R. (1993). Clarifying environmental education: A search for CLEAR ACTION in Southern Africa. *Southern African Journal of Environmental Education, 13*, 28–38.

Ramsarup, P. (2017). *A critical realist dialectical understanding of learning pathways associated with two scarce skill environmental occupations within a transitioning systems frame*. Unpublished PhD thesis, Rhodes University, Grahamstown, South Africa.

Ramsarup, P., Rosenberg, E., Lotz-Sisitka, H., & Jenkin, N. (2018). Chapter 8: Green skills: Transformative niches for greening work. In N. Mohamed (Ed), *Sustainability transitions in South Africa*, pp. 145–166. London: Earthscan/Routledge.

Raworth, K. (2017). *Doughnut economics: Seven ways to think like a 21st century economist*. London: Random House Business Books.

Rockström, J., Steffen, W., Noone, K., Persson, A., Chapin, F.S., Lambin, E.F., Lenton, T.M., Scheffer, M., Folke, C., Schellnhuber, H.J., Nykvist, B., de Wit, C.A., Hughes, T., van der Leeuw, S., Rodhe,H., Sörlin,S., Snyder, P.K., Costanza, R., Svedin, U., Falkenmark, M., Karlberg, L., Corell, R.W., Fabry, V.J., Hansen, J., Walker, B., Liverman, D., Richardson, K., Crutzen, P., & Foley, J.A. (2009). A safe operating space for humanity. *Nature*, *461*, 472–475, doi:10.1038/461472a.

Rotmans, J., & Kemp, R. (2003). *Managing societal transitions: Dilemmas and uncertainties: The Dutch energy case-study*. OECD Workshop on the Benefits of Climate Policy: Improving Information for Policy Makers, Paris.

Rotmans, J., Kemp, R., & van Asselt, M. (2001). More evolution than revolution: Transition management in public policy. *Foresight*, 3(1), 15–31.

Rotmans, J., & Loorbach, D. (2009). Complexity and transition management. *Journal of Industrial Ecology*, *13*,184–196.

SACMRM (The South African Coal Mining Road Map). (2013). *Fossil fuels, South Africa and SANEDI*, Johannesburg. Retrieved from www.fossilfuel.co.za and www.sanedi.org.za

Satgar, V. (Ed.). (2018). *The climate crisis: The South African and global democratic eco-socialist alternatives*. Johannesburg: Wits University Press.

Schwarz, S. (1994). Are there universal aspects in the structure and content of human values? *Journal of Social Issues*, *50*(4), 19–45.

Standing, G. (2010). *Work after globalization. Building occupational citizenship*. Cheltenham: Edward Elgar.

Statistics South Africa (StatsSA). (2019). Quarterly Labour Force Survey, Quarter 2: 2019. Statistical release PO211. Pretoria: Statistics South Africa.

Stern, M. (2007). *The economics of climate change: The Stern review*. Cambridge: Cambridge University Press.

Stern, N. (2018). Discussion following presentation at the OECD, Paris, December 2018.

Wahl, D.C. (2016). *Designing regenerative cultures*. Charmouth, Dorset and Chicago: Triarchy Press with International Futures Forum.

Ward, M. (2018). *Developing competencies for a just transition of the South African banking sector: Understanding just transition and its implications for the banking sector to inform a consideration of occupations and skills needs*. Johannesburg: Centre for Researching Education & Labour (REAL), University of Witwatersrand.

Weston, A. (1994). *Back to Earth: Tomorrow's environmentalism*. Philadelphia, PA: Templeton University Press.

World Bank. (2016). *World development indicators 2016*. Washington, DC: World Bank.

World Wide Fund for Nature (WWF). (2014). *Renewable energy Vision 2013 – South Africa*. Technical Report. Retrieved from http://awsassets.wwf.org.za/downloads.

World Wide Fund for Nature (WWF). (2017). *Food loss and waste facts and futures: Taking steps towards a more sustainable food future*. Cape Town: WWF-SA.

2 Green economy transitions and skills

Global and South African perspectives

Najma Mohamed and Presha Ramsarup

Introduction

In 2015, the 2030 Agenda for Sustainable Development and the Paris Agreement on Climate Change reaffirmed the centrality of and commitment to sustainable development in the global policy sphere. Sustainable development remains a conceptualisation which draws together a wide range of perspectives, analytic traditions and approaches (Scoones, Leach & Newell, 2015), converging in their commitment to the multi-dimensional objectives of achieving social inclusivity, environmental sustainability and economic prosperity.

There is large-scale consensus that climate and environmental threats are among the key global threats both in terms of impact and likelihood (World Economic Forum, 2019). Recent statements from the world's leading scientists have warned that urgent action is required to avoid the risks of extreme events, biodiversity loss and extreme heat and poverty for hundreds of millions of people (IPCC, 2014). This has led to concerns that climate change and its potential impacts could reverse the economic and social progress achieved globally and compromise the ability to reach many of the Sustainable Development Goals (SDGs).

Widespread and diverse actions characterise contemporary transformative shifts in policies, systems and institutions to achieve sustainable and climate resilient development pathways. These actions have emerged from 30–50 years' of concern for the state of the Planet. The Stockholm Conference on the Human Environment in 1972, followed by the Rio Earth Summit of 1992, the Johannesburg World Summit on Sustainable Development of 2002 and the Rio+20 Summit of 2012 have all raised these issues, and have been driving global consensus towards the SDGs. Today, transitions to inclusive greener economies are among the chief pathways adopted by countries for achieving sustainable development. This transition to greener economies is not only possible, but also affordable, and is among the key global trends affecting the future of work (International Labour Organisation [ILO], 2019). The gradual shift to low-carbon, greener economies is increasingly recognised and accepted as the context within which development will take place in the future.

In South Africa, the transition to a green economy is expressed as a "just transition to a resource-efficient, low-carbon and pro-employment growth path" in the *National Strategy for Sustainable Development* (NSSD) *and Action Plan* (Department of Environmental Affairs [DEA], 2011). In the NSSD, the transition to a green economy was identified as one of the five strategic sustainable development priorities for South Africa. A suite of supportive macro-economic and sector-specific policies now echo the government's commitment to a green economy transition conceptualised as a just and equitable transition to a low-carbon economy (Mohamed & Montmasson-Clair, 2019).

The challenge of persistent and growing unemployment has disproportionate effects on women and youth. Sluggish economic growth, owing to domestic and global market conditions, has resulted in annual economic growth rates far below the 5% growth rate required to make a dent in unemployment (Cotterill, 2019). Compounding these developmental challenges are the impact and challenges of climate change. Research confirms that South Africa is particularly vulnerable to climate change because of its dependence on climate-sensitive economic sectors such as agriculture, biodiversity, ecosystems and water, and high levels of poverty (Ziervogel et al., 2014; Vogel & Swilling, 2019).

Commitment to a low-carbon economy is captured in both the long-term vision of the *2020 New Growth Path* (Economic Development Department [EDD], 2009) as well as the *2030 National Development Plan* (NDP) (NPC, 2012), which expressly identifies the green economy as one of the key sectors for investment and job creation. The creation of green and decent jobs is at the core of South Africa's green economy vision. The release of a national *Green Jobs* report in 2011 (Maia et al., 2011) further confirmed the employment creation potential of an unfolding green economy. It was estimated that a transition to a green economy could lead "to the creation of approximately 98 000 new direct jobs, on average, in the short term, almost 225 000 in the medium term and around 462 000 employment opportunities in the formal economy in the long term" (Maia et al., 2011, p. 4). Skills shortages were highlighted among the key implementation challenges facing the creation of green jobs. The development of a coherent skills development strategy, including worker reskilling, was identified as key to the 'greening' of economic sectors.

"Investment in skills, technology and institutional capacity is critical to support the development of a more sustainable society and the transition to a low-carbon economy", according to South Africa's *National Development Plan* (NPC, 2012, p. 197). Starting with a short history of sustainable development and green skills research emerging as two interconnected 'tales' of transition (Ramsarup, 2017), this chapter will then focus in more depth on green economy transitions, seeking to outline the context for the emergence of more recent green skills policies, models and practices both globally and in South Africa.

A brief history of sustainable development and green skills emergence in South Africa

In 1996 the post-apartheid government introduced a people-centred sustainable development framework for environmental management and conservation into the South African policy landscape. Environmental rights and sustainable management of South Africa's natural resources were included in the Bill of Rights in Section 24 of the Constitution. Environmental policy in South Africa was "underscored by an equity- and rights-based discourse in which sustainable development visions include a strong emphasis on social justice" (Mohamed, 2019, p. 5).

The Department of Environmental Affairs and Tourism, as lead agent for environmental management, promulgated the National Environmental Management Act in 1998 to give effect to the Constitution. Since then, a series of associated legislation and policy has been promulgated, each with specific skills related needs. The environmental sector in South Africa, as envisaged by the Constitution, can therefore be described as a new, and *rapidly emerging* and *developing sector* (see Figure 2.1).

The environmental policy formulation processes placed parallel demands for transformation of the education, training and skills sector, and the emergence of green skills in South Africa reflects a co-evolving relationship between two major systems, both of which have been in transition, with overlaps and disjunctures as depicted in Figure 2.1. For example, when the Constitutional clause was promulgated in 1996, and the National Environmental Management Act of 1998 was proclaimed, the first White Paper on Education and Training of 1995, which included a principle on environmental education as a cross-cutting concern for all levels and phases of the education and training system, was released. At this stage, there were only a few environmental qualifications or specialist environmental science, environmental education, environmental law, environmental sociology, environmental engineering, etc. courses available in South Africa, due to this not being a major priority of the apartheid state, especially from a human rights, social justice and sustainable development perspective.

With increasing focus on sustainable development, green economy and just transitions in the South African environmental policy, there have been some landmark education and skills system responses: The environment was included as a cross-cutting concern in the national curriculum from as early as 1996, new environmental qualifications were in place in the early 2000s and by 2010 South Africa developed its first *Environmental Sector Skills Plan* (ESSP) (DEA, 2010). The ESSP developed out of a realisation that the education, training and skills policy and practice system was not keeping 'pace' with the rapid emergence of new environmental policy and discourse.

As much as green skills development is meant to support the emerging discourse on green economies and just transitions, i.e. it is integral to economic sectors and the greening of work (see below), the environmental sector also has a public

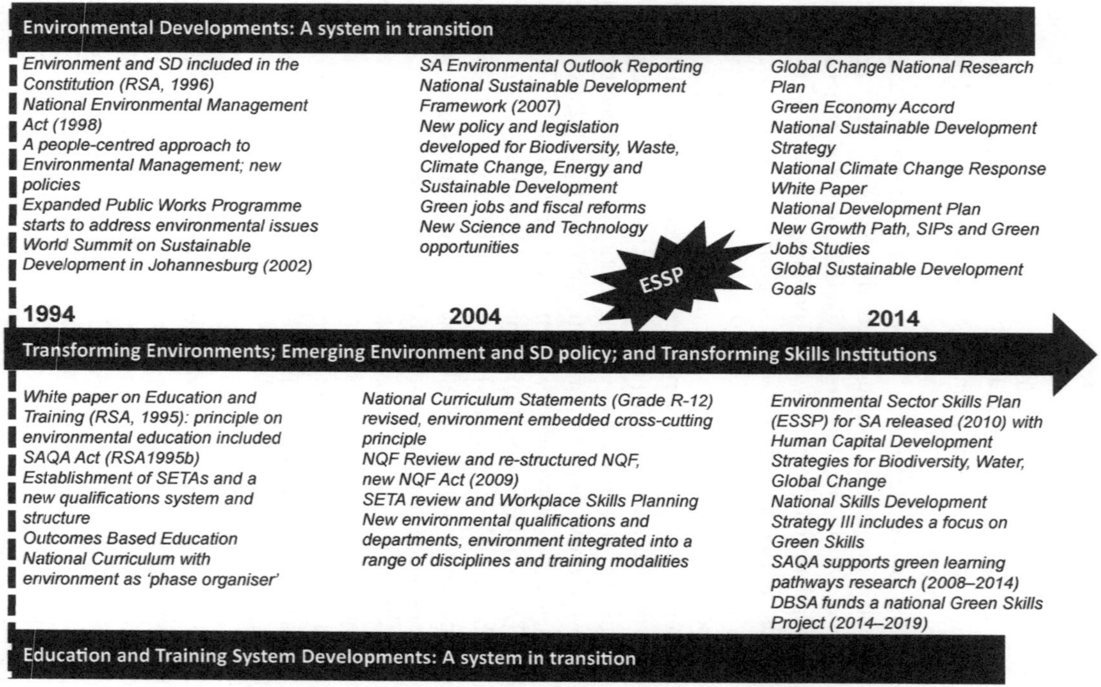

Figure 2.1 Environment and sustainable development and education, training and skills development systems in interconnected transition

Source: Adapted from Lotz-Sisitka, Ramsarup and Mosidi, 2010 and Ramsarup, 2017 good mandate. It carries the responsibility of ensuring that South Africa's natural resources and ecosystem services are to be sustainably used and managed for present and future generations. This dual focus has affected the overall emphasis given to the environmental skills supply in South Africa, and by 2010, there was a crisis in skills provisioning for the burgeoning environmental sector, and the Department of Environmental Affairs commissioned the Environmental Sector Skills Plan (DEA, 2010), raising the need for a systemic, more proactive approach to human capacity development in the environmental sector. Figure 2.1 provides an illustration of these two interrelated systems, both in transition.

One of the complexities of developing environment and sustainable development capacity and skills is their cross-sectoral, cross-cutting nature. Adequate forums and institutional structures are required to support skills planning for cross-cutting concerns in a transversal way, an issue that the South African Qualifications Authority (SAQA) sought to address by supporting a sustainable development workplace learning and transversal green skills learning pathways research programme in 2008 (Lotz-Sisitka & Ramsarup, 2017). This SAQA research programme, together with the research undertaken for the ESSP and a Biodiversity Human Capital Development Strategy (Human Sciences Research Council [HSRC], 2009) identified the lack of adequate research frameworks for green skills research and planning in South Africa. Rosenberg, Lotz-Sisitka and Ramsarup (2018) provided one example of a learning framework that explored the learning needs for green economy policy practitioners (cf. Chapter 9).

Green economy transitions: diverse and divergent

Despite being coined more than 20 years ago, the term 'green economy' has only recently found its way, in various permutations, into the prevailing public policy discourse on sustainable development across the world, including South Africa (also shown in Figure 2.1). Although the green economy emerged as one of the four central themes at the 2012 United Nations Conference on Sustainable Development (Rio+20), this economic transformation is characterised by the varied approaches, visions, models and tools available to each country, in accordance with its national circumstances and priorities, to achieve sustainable development (United Nations, 2012).

Green economic development is promoted by a wide range of actors and has even "reached the core of global economic elites, which represents a change of paradigm that environmental scientists and environmentalists have been calling for for decades" (Jänicke, 2012, p. 13). In a seminal publication, *Towards a Green Economy*, the United Nations Environmental Programme (UNEP, 2011) outlined some of the cogent benefits that could emerge from a green economy transition including: an increase in wealth, encompassing gains in both natural capital and a higher rate of GDP growth; better maintenance and conservation of ecosystems which will support efforts towards poverty eradication; and the creation of green jobs. Diverse and divergent emphases in green economy discourses exist (cf. also Chapter 12), reflecting various concerns. It is possible to identify the distinguishing features of the various discourses around greening the economy such as a focus on economic growth, ecosystem resilience, job creation and a focus on low-carbon economies or low-emission development through the reduction of greenhouse gases.

Critical voices, pointing to the neglect of the social dimensions of prevailing green economy discourses, highlight the fact that some of the proposed "economic, technological and institutional changes that currently form the basis of green economy strategies run the risk of reinforcing human insecurities and inequalities" (United Nations Research Institute for Social Development [UNRISD], 2012). The growing body of work exploring the integration of social justice in sustainable development policymaking (Leach, Scoones & Stirling, 2010; IIED & CAFOD, 2014; King, Olhoff & Urama, 2014; UNRISD, 2016) highlights the need for more transformative green economy pathways. If the current trend of sustainable development policies is to represent a transformative eco-social and pro-poor turn well-being, social justice, governance, equity and redistribution have to be integrated in the policy cycle in a more meaningful way.

There are several theoretical framings of green economy discourses that can be used to disaggregate emerging debates and understand the multiplicity of pathways towards greening economies (Jacobs, 2012; Death, 2014; Faccer, Nahman & Audouin, 2014; cf. also Chapters 12 and 14). The key benefits of developing a conceptual map and understanding of diverse green economy discourses is to assist in "understanding the origins, as well as the advantages and disadvantages, of particular 'green economic' strategies, and to select those most appropriate within particular circumstances" (Faccer et al., 2014, p. 645). Table 2.1 provides a comparison of two studies which present discourses on the green economy.

Globally, the various green economy transition discourses can be discerned in the positions of key proponents. In South Africa, tensions and contradictions between technocentric and more radical approaches, which call for structural transformation, can also be discerned in the embryonic green economy discourses. It can be deduced from the concerted efforts by powerful vested interests, allied to the state, to derail the transition towards renewable energies in South Africa, or in the ways in which debates between proponents of natural capital have trumped discourses which promote the application of non-monetary indices in the valuation of nature (Mohamed & Montmasson-Clair, 2019).

Death (2014), for instance, acknowledged that all the discourses outlined in Table 2.1 can be discerned in invocations of the green economy in South Africa, but the dominant discourse of the green economy in the country is regarded as that of green growth. It is by capitalising on

> South Africa's 'brand' as a rising power, with a youthful and energetic population and a rich natural environment, and the country as the political and economic 'gateway to Africa', rather than its environmental and social contradictions, that South Africa can be positioned as a global leader on the green economy.
>
> (Death, 2014, p. 2)

Table 2.1 Key features of two discourse framings on the green economy

Four discourses of the green economy (Death, 2014)	*Discourses related to the green economy (Faccer et al., 2014)*
Green Revolution: radical, revolutionary transformation on economic (and hence social and political) relationships to bring them in line with natural limits and ecological virtues	*Transformative Discourse*: incorporates critical perspectives calling for a more radical review of society's economic and broader developmental objectives
Green Transformation: explicit focus on social justice, equity and redistribution (including intergenerationally) where economic growth is a means rather than an end	*Reformist Discourse*: diverse agendas for a green economy, with an emphasis on the right combination of actions and long-term planning to achieve environmental benefits as well as stronger economic growth
Green Growth: green markets provide economic opportunities representing a recasting of the relationship between environment and economics with an emphasis on new markets, new services and new forms of consumption	*Incrementalist Discourse*: defined by a broad acceptance of the prevailing macro-economic paradigm and a focus on greater use of market-based tools to drive a green economy transition
Green Resilience: essentially reactionary and cautious with an emphasis on environmental scarcity, climate change and resource depletion and the need to implement technological solutions to build local self-sufficiency/resilience	

While there are voices articulating green transformation and green revolution transition discourses in South Africa – agitating for green and decent jobs, just transitions, food sovereignty, climate justice and energy democracy – these have not coalesced into a coherent movement which could present a counterforce to the green growth discourse. The language of the green economy, notably the varieties which centralise equity, was initially welcomed by environmentalists in South Africa for its focus on social and economic inclusion.

Among these alternative and competing green economy discourses, the technocentric, market-led discourse of green growth is most embedded and widespread in South Africa, with limited intersection with social justice, rights and redistributive discourses central to more inclusive approaches. This is beginning to change as innovations in policies and practices, including those related to building skills for a greener economy, emerge.

South Africa's transition to a green economy: key drivers

South Africa's transition to a green economy is embedded within its macroeconomic policy framework and features prominently in the long-term development plan of the country. The foundational thinking for this transition has its antecedents in the country's response to the 2009 global financial crisis; the employment-generating potential of green economy transitions ('the promise of green jobs') and its commitment to the reduction of greenhouse gas emissions. The key drivers for the country's green transition, including some of the challenges the country faces in transitioning to a green economy, are outlined in this section.

Political drivers

In 2012, South Africa was ranked first in global leadership in the Global Green Economy Index due to the strength of its policy and political commitment towards greening its economy (Dual Citizen, 2012). This followed the hosting of the 17th Conference of Parties (COP 17) in 2011, which acted as the fulcrum for deliberations on global climate change commitments. By 2016, South Africa's performance had declined notably, losing the impetus created by hosting this prestigious global environmental event. The country was placed 59th among the 80 countries covered in the Index (Dual Citizen, 2016). Further, Yale's Environmental Performance Index in 2016, which ranks countries in relation to their performance on high priority environmental issues in two areas: protection of human health and protection of ecosystems, placed the country 81 out of the 90 countries measured (Yale Centre for Environment and Policy, 2016).

A multitude of political challenges including widespread corruption, which spanned almost a decade, saw dramatic political changes in South Africa with the resignation of President Jacob Zuma in 2018. In 2019, the new presidency heralded the 'dawn of a new era', proposing to address the political and development challenges of the country through innovation- and investment-led growth, employment and transformation. This innovation-led growth strategy supports economic growth via investment in green industries and technologies. A 2016 analysis of South Africa's green economy landscape identified close to 1 000 active 'green economy' initiatives in the country, affirming that "technological innovation towards low-carbon and resource-efficient technologies has had significant uptake in South Africa" (Partnership for Action on Green Economy [PAGE], 2017, p. 75). In reality though, the country continues to walk the tightrope between a coal-based economy and implementing the steps needed to transition towards a green economy.

Policy and regulatory drivers

Emerging from the first national Green Economy Summit in 2010, South Africa enacted a suite of supportive policies, strategies and incentives confirming the country's commitment to green economy objectives. This was expressed in the short, medium and long-term vision of the country, outlined in the *National Strategy for Sustainable Development and Action Plan* (2014), *New Growth Path* (2020) and *National Development Plan* (2030). Strategic policy documents such as the *Environmental Fiscal Policy Reform Paper* (National Treasury, 2006), *10-Year Innovation Plan* (Department of Science and Technology [DST],

2008), *Green Economy Accord* (EDD, 2011) and *Industrial Policy Action Plan II* (Department of Trade and Industry [DTI], 2010), which have an economy-wide impact, have also been enacted. Following the *National Climate Change Response White Paper* of 2010, the Department of Environmental Affairs has released the long-awaited *Climate Change Bill* (DEA, 2018), focused largely on the institutional and governance framework required to formulate an effective climate change response.

The Green Economy Accord, formulated in the context of South Africa's New Growth Path, launched a partnership between government, business, trade unions and civil society. It identified the green economy as one of the ten priority growth areas for the country, and set goals to create 300 000 new green jobs in green industries in contribution to the *New Growth Path's* objective of creating five million new jobs by 2020. The macro-economic policy framework, which cuts across economic sectors, is complemented by a suite of sector-specific policies which reinforce the country's commitment to a green economy transition. For instance, the energy, transport, waste management and construction sectors have seen the enactment of supportive green economy policies.

Applying the policy categorisation of King et al. (2014) in Figure 2.2, green economy policymaking in South Africa has largely been within the ambit of developing enabling policies, both at the macro-economic and sector-specific level. Mandating and/or incentivising policies, which would deepen policy impacts, such as the environmental fiscal reform process initiated in 2006 or the tax incentives for the implementation of energy efficiency measures, remain limited. The need for a coherent and consistent policy framework, which not

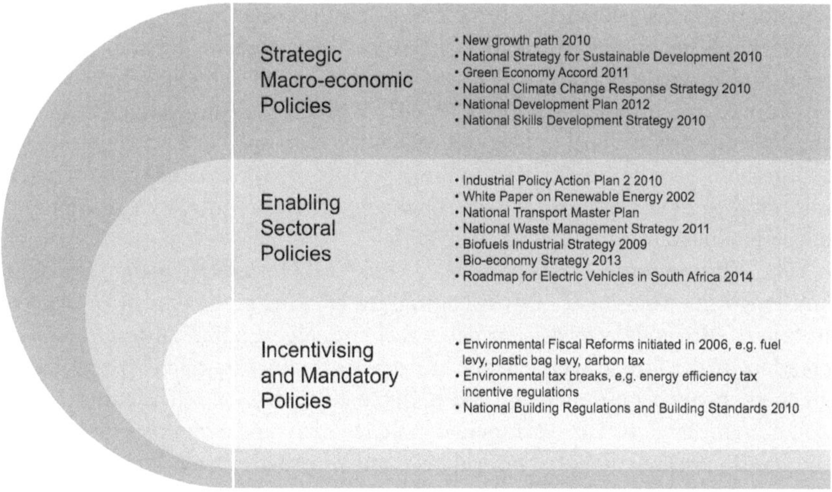

Figure 2.2 Key policy drivers supporting the transition to a low-carbon economy in South Africa

only provides clear policy signals, but extends and deepens policy impact through incentivising and mandating policy instruments, is needed to strengthen green economy actions.

Socio-economic drivers

The social and economic benefit of transitioning to a green economy has been one of the main drivers of the green economy transition in South Africa, framed as a 'just transition'. Swilling, Musango and Wakeford (2015) posited that in South Africa "a just transition can be understood as a structural transformation that results in the achievement of two linked goals: developmental welfarism and a sustainability transition" (p. 2). Despite gains in economic development since its transition to democracy, the country continues to face the triple challenges of rising unemployment, income inequality and poverty. The *Framework for South Africa's Response to the International Economic Crisis* (2009), crafted by government, business and labour, sought to develop (and invest in) technologies and industries that could combat the negative effects of climate change. This was identified as an important fiscal stimulus that would drive both economic growth and employment creation. This vision was carried forward in subsequent policies, such as the New Growth Path and the Green Economy Accord which set particular targets for the creation of green jobs.

Amid a volatile global economy, South Africa continues to experience slow economic growth and steadily increasing unemployment rates. In fact, "the country is plagued by one of the highest levels of unemployment and economic inactivity for middle-income countries" as "more than one-third of its labour force is out of work or has stopped searching for jobs" (World Bank, 2015). Poverty and inequality levels remain high. However, the Organisation for Economic Co-operation and Development's (OECD's) 2013 *Economic Survey of South Africa* states that "effective green growth policies should be combined with other structural and macro-economic policies to reconcile rapid economic growth with environmental sustainability" (OECD, 2013a, p. 91). Continued subsidisation of the carbon-intensive coal industry, as well as the slow uptake of pricing strategies to internalise environmental externalities such as the (postponed) carbon tax, is impeding economic reforms which could support the country's green economy development trajectory (OECD, 2015).

The 2011 report on green jobs (Maia et al., 2011) presented a short, medium and long-term assessment of the potential for creating green jobs in key sectors in South Africa. The report estimates that the largest potential for green jobs creation in the formal sector lies in the areas of natural resource management, green energy generation, energy and resource efficiency, as well as emission and pollution mitigation. The enactment of policies, plans and incentives to stimulate the creation of green jobs has not yet been implemented at scale, with the exception of renewable energy. There are promising examples emerging from green jobs initiatives in public works, waste management, ecological maintenance and transport. Similarly, the advancement of the socio-economic benefits of green investments (beyond job creation), such as the promotion of youth employment, gender equality and the informal economy, has thus far been marginal in large-scale green economy projects. There are also concerns that

> South Africa's planning blueprint, the National Development Plan, uses the language of a just transition but inflects it towards a green neoliberal approach, in which market mechanisms are used to mitigate greenhouse gas emissions, promote technofixes (such as carbon capture and storage), incentivise energy efficiency in business, and support the expansion of fossil fuel extractivism by mining coal and shale gas.
>
> (Satgar, 2014, p. 148)

A just transition that extends beyond incrementalist discourses on green economic transformation towards reformist and transformational discourses that are unequivocally committed towards addressing the environmental *and* social dimensions of sustainability transitions is thus needed (cf. also Chapter 1).

Climate change and environmental drivers

South Africa has developed a comprehensive policy and regulatory framework for environmental and natural resource management, and established a cooperative governance regime to support the implementation of climate and environmental policies. Noted for its natural wealth, the country is ranked among the most biologically diverse countries in the world. However, the OECD's *Environmental Performance Review: South Africa* (2013b) found that environmental challenges such as climate change and water scarcity threaten the sustainability of economic growth in South Africa. The climate constraints facing this carbon-intensive economy are one of the principle drivers of the country's commitment towards a low-carbon development path. The National Development Plan (National Planning Commission [NPC], 2012, p. 209) states that the country's "primary approach to adapting to climate change is to strengthen the nation's economic and societal resilience" and measures to support this policy vision has included market mechanisms (e.g. carbon taxes), regulatory mechanisms (e.g. energy efficiency standards) and societal engagements (e.g. the voluntary Carbon Disclosure Project initiated by business).

In 2009, the country committed itself to reducing greenhouse gas (GHG) emissions through a "peak, plateau and decline" in its emissions. Recognising the need to shift the economy's growth path, the government is committed to reducing GHG emissions by 34% by 2020 and by 42% by 2025 on condition that it receives technology and finance support from developing countries. This is a difficult challenge because the country is one of the most energy-intensive economies in the world, and is a significant contributor to global GHG emissions owing to its fossil-fuel powered economy. Further, as a water-stressed country, it is extremely vulnerable to the impacts of climate change. Ziervogel et al. stated that

> climate change poses a significant threat to South Africa's water resources, food security, health, infrastructure, as well as its ecosystem services and biodiversity. Considering South Africa's high levels of poverty and inequality, these impacts pose critical challenges for national development.
>
> (2014, p. 606)

Climate change and environmental drivers have been at the forefront of making the case for transitioning to a green economy. In last 50 years, South Africa has already experienced temperature increases 1.5 times those of the global level and climate change impact projections predict warming relative to 1986–2005 of 3–6°C by 2081–2100 in the interior of the country (Ziervogel, et al., 2014). A vision for an effective climate change response and long-term transition to a climate-resilient, efficient and equitable low-carbon economy and society exists, but will require effective and timely coordination to bridge the chasm between policy and practice.

Technological drivers

South Africa has developed an enabling policy framework for developing green technologies. However, the country scored well below average in the Global Cleantech Innovation Index (2014) which looks at the ability of countries to create the right environment for green entrepreneurial activity. An assessment commissioned by the Department of Science and Technology, *The State of Green Technologies in South Africa* (ASSAf, 2014), provides a useful overview of the green technology landscape. The study found that while there is a strong emphasis on green technologies in the energy sector, the uptake of green technologies in other sectors is generally slow. This is not surprising given the country's track record of technology uptake which has generally been below average. Significant barriers to the implementation of green technologies, such as a lack of policy coherence and regulatory certainty to support the development of green technologies, government bureaucracy, skills shortages and access to finance, still exist. Renewable energy and energy-efficient technologies and green technological innovations in the waste, water and construction sectors point to sectors in which technological innovations have been implemented at a faster rate than others (ASSAf, 2014).

The NDP notes that investment in skills, technology and institutional capacity is critical to support the development of a more sustainable society and the transition to a low-carbon economy. There is an increasing effort to grow and support climate innovations and to provide business support development and finance to small and medium size enterprises which are operating in the green economy space. This will have to be scaled-up considerably to harness technological innovation and green enterprise development in support of the country's green economy transition.

South Africa has instituted several enabling conditions to support its vision of ensuring environmental sustainability and an equitable transition to a green economy, yet it still faces several challenges. The energy and carbon-intensity of its economy poses a critical challenge to instituting pricing and technological reforms, which could diversify the energy mix. Even as the country moves rapidly towards meeting targets for increasing the renewable energy component in its energy mix, significant challenges remain to implement reforms in the energy sector, notably the implicit and explicit subsidisation of the fossil fuel industry (Montmasson-Clair, 2017). Despite the country's commitment to a green economy transition, implementation of green economy programmes and initiatives at scale is hampered by policy incoherence, regulatory uncertainty and a lack of coordination of green economy activities in the country. This is compounded by a limited human capacity to design, implement, manage and monitor green economy policies and initiatives (PAGE, 2016).

The *National Skills Development Strategy III* (Department of Higher Education and Training [DHET], 2011) of South Africa spells out the need to build institutional and organisational capacity to meet the labour and skills demands of a green economy; yet skills shortages in the country remain critical. In 2010, a country study on skills for green jobs found that there were "significant skills gaps across all sectors" that would hamper the development of a low-carbon economy in South Africa (ILO, 2010), confirming the DEA (2010) finding that the skills system was largely re-active to environmental concerns. Globally, skills mismatches have been identified as one of the major barriers to the greening of economies (ILO, 2018), while skills are central to unlocking the job creation opportunities arising from green transitions.

Skills for a green economy

Skills can be a key enabling factor in green transitions and are beginning to feature prominently in green economy strategies and plans (ILO, 2011a; OECD, 2013c). Green skills identification, anticipation and planning are vital to extract the full employment-creation benefits from green economy transitions. This requires integration of green skills development in national skills planning systems, alignment of skills development with environmental policies and coordination between employers, policymakers, educational institutions and labour. Reflecting the systemic and integrated nature of green economy transitions,

> green skills can be defined as the skills needed by the workforce, in all sectors and at all levels, in order to help the adaptation of products, services and processes to the changes due to climate change and to environmental requirements and regulations.
>
> (OECD, 2013c, p. 10)

Identified as one of the four key trends that impact on the future of work (ILO, 2019), the greening of economies will result in the creation of millions of jobs, but also in some job losses as economies decarbonise. According to the 2018 ILO study, *Greening through jobs*, implementing the Paris Climate Change Agreement will lead to global job losses of around 6 million and job gains of 24 million. The European Centre for the Development of Vocational Training (Cedefop) (2012) has outlined three ways in which green economy transitions could affect skills. First, structural changes could occur in sectors, such as the energy, mining and manufacturing sectors, which will lead to an increase in demand for some occupations and a decline in demand for others. A second impact could be on the creation of new occupations which will require the development of new skill profiles, qualifications and training programmes. And thirdly, existing occupations will have to be 'greened' and adjustments made to existing training and qualification frameworks (retraining or upskilling) for these occupations (cf. Chapter 12).

Despite the compelling case for green skills planning, to meet the demand for these skills but also to drive green economy transitions, there are several systemic challenges in green skills development. These include: a lack of structures to deal with green skills as a cross-cutting concern in the education and training landscape, a lack of education and training opportunities and dated curricula in training and education institutions, inadequate labour market analysis instruments for capturing green occupations and a discord between skills and green economy policies, as evident across the chapters of this book.

South Africa is among the countries which display a weak coherence between environmental and skills policies despite advancements in its national policy framework on sustainable development (see also DEA, 2010). While this discord remains a challenge in many countries, there are promising examples emerging, notably at a sectoral level (and in energy in particular), where skills for green transformation are emerging as a priority. Progress in skills formulation for green transitions are more visible at sectoral level where regulatory instruments have been particularly effective in setting green skills formulations in motion (ILO, 2018). Particular attention has been focused thus far on either skilled or semi-skilled jobs (notably in energy), while the requisite support for the identification of green skills for low-skilled workers has not been forthcoming (ILO 2018; cf. also Chapter 8). This is particularly important in many developing country contexts where there are still large numbers of low-skilled workers as well as workers operating in the informal economy. If the creation of green and decent jobs is to be prioritised, then active labour market policies that target vulnerable workers need to be implemented (as argued in Chapter 8), and relevant approaches to skills development at this level also need to be developed, which requires research to better define green occupations and job roles at this level (cf. Chapter 12).

The ILO (2011b) identified four levels of skills analysis: macro-economic, sectoral, occupations and skills, training and education. The level of analysis and selection of methodologies to undertake an analysis of the green skills requirements will be driven by the research questions, as well as the availability of data and research capacity. The ILO (2011c) guideline document entitled, *Comparative Analysis of Methods of Identification of Skill Needs on the Labour Market in Transition to the Low Carbon Economy*, seeks to align the level of analysis with the appropriate methodology and headline research questions. Often though, as is the case in South Africa, green skills studies span several levels of analysis – with a strong focus on the sectoral level, a point made across this book through the argument for adopting a laminated systems approach to skills research. Figure 2.3 highlights the ILO typology of the levels of analysis in skills research with reference to green skills studies undertaken or underway in South Africa (cf. also Chapter 14).

Green skills often span more than one level as the reports in South Africa illustrate. For instance, both the *Skills for Green Jobs in South Africa* (ILO, 2010) and the report with green job projections (Maia at al., 2011), which have adopted quantitative modelling methods to address the macro-economic implications of green transitions, incorporate sector-level analyses. Sectoral green skills studies, in mining and biodiversity for instance, also incorporate skills and occupation-level analysis (cf. Chapters 3, 4 and 5). The *Skills for and through SIPs* (EDD and DHET, 2014) study has adopted a multi-level analysis which has centred on the occupation and skills-level. This skills study has included the identification of

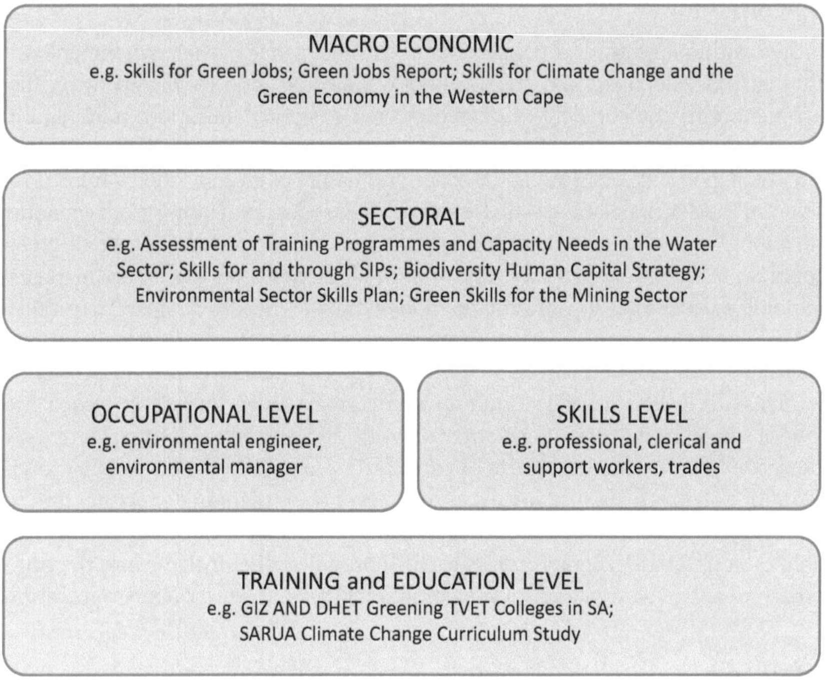

Figure 2.3 Overview of different levels of analysis in skills research with examples from South Africa

Source: Adapted from ILO 2011b

the critical occupations needed to deliver strategic infrastructure projects, such as Environmental Engineers and Environmental Managers, and has established occupational teams to further identify education and training requirements for developing these skills (cf. also Chapters 9 and 12). Training and education-level analyses, undertaken in the technical and vocational education training (TVET) and university (Integrating Climate Change in University Curricula) sectors, have looked at the training of trainers, curriculum content and structure and training methodologies (cf. also Chapters 10 and 11).

The 2010 *Skills for Green Jobs* and 2011 *Green Jobs Report* in South Africa both allude to the need for a cohesive approach to green skills development at a national level. However, mainstreaming green skills development within the national skills planning system remains a challenge, as elaborated in Chapter 14 of this book. Efforts around green skills planning, research and development, like many of the key enabling processes required for South Africa's green economy transition, such as the coordination and streamlining of environmental finance, remain ad hoc and disparate (Mohamed, 2019). However, promising examples, such as the models, methods and cases presented in this book, point to the growing body of work which interrogates skills development in green transitions.

Conclusion

Located within an historical perspective on the emergence of the environment and sustainable development and green economy trajectory, and its alignment or misalignment with the parallel yet interconnected emerging education, training and skills development system in South Africa, this chapter sought to reaffirm and highlight the importance of assessing the scope and nature of green economy transitions to inform the development of skills for sustainability. Contextualising green economy transitions, by analysing macro-economic drivers, illustrates the impact of green economic restructuring of sectors and industrial segments on the labour market – including the identification and anticipation of the skills needed for green transitions.

The impact of green economy strategies and plans on skills development will differ considerably, even within a country, given the focus on different economic sectors, skills levels and education and training capacities. Further, the pace and scale of an economy-wide transformation will differ between sectors, as already observed. While a national system of green skills identification and anticipation is required to mainstream green skills in high-level skills planning structures and in macro-level skills development strategies, context-specific studies, whether sector-based or at different geographical scales, will provide richer insights into the skills and attendant education and training requirements of green economy transitions.

References

Academy of Science of South Africa (ASSAf). (2014). *The state of green technologies in South Africa*. Pretoria: ASSAf.

Cleantech Group and WWF. (2014). The Global Cleantech Innovation Index 2014. Cleantech Group and WWF.

Cotterill, J. (2019). South Africa's economic growth stutters. *Financial Times*, 9 March 2019. Retrieved from www.ft.com/content/1688aa70-3f53-11e9-b896-fe36ec32aece

Death, C. (2014). The green economy in South Africa: Global discourses and local politics. *Politikon: South African Journal of Political Studies*, *41*(1), 1–22.

Department of Environmental Affairs (DEA). (2010). *Environmental Sector Skills Plan for South Africa*. Pretoria: DEA.

Department of Environmental Affairs (DEA). (2011). *National Strategy for Sustainable Development and Action Plan*. Pretoria: DEA.

Department of Environmental Affairs (DEA). (2018). *National Climate Change Bill*. Pretoria: DEA.

Department of Higher Education and Training (DHET). (2011). *National Skills Development Strategy III*. Pretoria: DHET.

Department of Science and Technology (DST). (2008). *Ten-year Innovation Plan*. Pretoria: DST.

Department of Trade and Industry (DTI). (2010). *Industrial Policy Action Plan II*. Pretoria: DTI.

Dual Citizen. (2012). *The 2012 Global Green Economy Index Report*. Washington, DC: Dual Citizen.

Dual Citizen. (2016). *Global Green Economy Index: 2016*. Washington, DC: Dual Citizen.

Economic Development Department (EDD). (2009). *New Growth Path*. Pretoria: EDD.

Economic Development Department (EDD). (2011). *Green Economy Accord*. Pretoria: EDD.

Economic Development Department (EDD) & Department of Higher Education & Training (DHET). (2014). *Skills for and through SIPs. What has been dome and still needs to be done to skill South Africans for SIPs and through SIPs*. Pretoria: EDD and DHET.

European Centre for the Development of Vocational Training (Cedefop). (2012). *Skills for green jobs*. Briefing Note. Thessaloniki: Cedefop.

Faccer, K., Nahman, A., & Audouin, M. (2014). Interpreting the green economy: Emerging discourses and their considerations for the Global South. *Development Southern Africa*, *31*(5), 642–657.

Global Green Growth Institute (GGGI). (2016). *Pro-poor, inclusive green growth: Experience and new agenda*. Seoul: GGGI.

Human Sciences Research Council (HSRC). (2009). *Guidelines towards a human capital development strategy in the biodiversity sector*. Research report: Foundation and the South African National Biodiversity Institute (SANBI), October 2009. Pretoria: HSRC.

International Institute for Environment and Development (IIED) and Catholic Agency for Overseas Development (CAFOD). (2014). *Securing social justice in green economies*. London: IIED.

International Labour Organisation (ILO). (2010). *Skills for green jobs in South Africa*. Unedited background country study. ILO Skills and Employability Department. Retrieved from www.ilo.org/skills/what/projects/lang--en/WCMS_115959/index.htm

International Labour Organisation (ILO). (2011a). *Skills for green jobs: A global view. Synthesis report based on 21 country studies*. Geneva: ILO.

International Labour Organisation (ILO). (2011b). *Anticipating skills needs for the low carbon economy? Difficult, but not impossible*. Research Brief. Geneva: ILO.

International Labour Organisation (ILO). (2011c). *Comparative analysis of methods of identification of skill needs on the labour market in transition to the low carbon economy*. Geneva: ILO.

International Labour Organisation (ILO). (2018). *World employment and social outlook 2018: Greening with jobs*. Geneva: ILO.

International Labour Organisation (ILO). (2019). *Work for a brighter future: Report of the Global Commission on the Future of Work.* Geneva: ILO.

Intergovernmental Panel on Climate Change (IPCC). (2014). *Climate change 2014: Synthesis report.* Contribution of Working Groups I, II and III to the Fifth Assessment Report of the Intergovernmental Panel on Climate Change [Core Writing Team, R.K. Pachauri and L.A. Meyer (Eds.)]. Bonn: IPCC.

Jacobs, M. (2012). *Green growth: Economic theory and Political Discourse Centre for Climate Change Economics and Policy.* Working Paper No. 108, Grantham Research Institute on Climate Change and the Environment Working Paper No. 92.

Jänicke, M. (2012). Green growth: From a growing eco-industry to economic sustainability. *Energy Policy*, 48, 13–21.

King, P., Olhoff, A., & Urama, K. (2014). Policy design and implementation. In *Green growth in practice: Lessons from country experiences.* Seoul: GGBP.

Leach, M., Scoones, I., & Stirling, A. (2010). *Dynamic sustainabilities. Technology, environment, social justice.* Abingdon: Earthscan.

Lotz-Sisitka, H.B., & Ramsarup, P. (2017). Learning pathways and articulation: Early conceptual explorations and implications for research design(s). *SAQA Bulletin: Learning pathways, the National Qualifications Framework (NQF), and lifelong learning in South Africa, 17*(1), 31–48.

Maia, J., Giordano, T., Kelder, N., Bardien, G., Bodibe, M., Du Plooy, P., Jafta, X., Jarvis, D., Kruger-Cloete, E., Kuhn, G., Lepelle, R., Makaulule, L., Mosoma, K., Neoh, S., Netshitomboni, N., Ngozo, T., & Swanepoel, J. (2011). *Green jobs: An estimate of the direct employment potential of a greening South African economy.* Industrial Development Corporation (IDC), Development Bank of Southern Africa (DBSA) and Trade and Industrial Policy Strategies (TIPS), Pretoria.

Mohamed, N. (2019). *Sustainability transitions in South Africa.* London: Routledge.

Mohamed, N., & Montmasson-Clair, G. (2019). Policies for sustainability transformations in South Africa: A critical review. In N. Mohamed, *Sustainability transitions in South Africa.* London: Routledge.

Montmasson-Clair, G. (2017). *Electricity supply in South Africa: Path dependency or decarbonisation?* TIPS Policy Brief 02/2017. Pretoria: Trade & Industrial Policy Strategies.

National Planning Commission (NPC). (2012). *National Development Plan 2030.* Pretoria: NPC.

National Treasury. (2006). *A Framework for Considering Market-Based Instruments to Support Environmental Fiscal Reform in South Africa.* Draft Policy Paper. Pretoria: National Treasury, Tax Policy Chief Directorate.

Organisation for Economic Co-operation and Development (OECD). (2013a). *Economic survey of South Africa 2013.* Paris: OECD.

Organisation for Economic Co-operation and Development (OECD). (2013b). *Environmental performance review: South Africa.* Paris: OECD.

Organisation for Economic Co-operation and Development (OECD). (2013c). *Greener skills and jobs for a low-carbon future.* Working Papers, 2013-10. Paris: OECD.

Organisation for Economic Co-operation and Development (OECD). (2015). *Economic Survey of South Africa 2015.* Paris: OECD.

Partnership for Action on Green Economy (PAGE). (2016). *Green economy learning assessment South Africa: Critical competencies for driving a green transition.* Geneva: PAGE.

Partnership for Action on Green Economy (PAGE). (2017). *Green economy inventory for South Africa: An overview.* Geneva: PAGE.

Ramsarup, P. (2017). *A critical realist dialectical understanding of learning pathways associated with two scarce skill environmental occupations within a transitioning systems frame*. Unpublished PhD thesis, Rhodes University, Grahamstown, South Africa.
Rosenberg, E., Lotz-Sisitka, H. B. & Ramsarup, P. (2018). The green economy learning assessment South Africa: Lessons for higher education, skills and work-based learning. *Higher Education, Skills and Work-Based Learning*, 8(3), 243–258.
Satgar, V. (2014). South Africa's emergent green developmental state. In M. Williams (Ed.) *The End of the Developmental State?*. Pietermaritzburg: UKZN Press.
Scoones, I., Leach, M., & Newell, P. (2015). *The politics of green transformations*. Abingdon: Routledge.
Swilling, M., Musango, J., & Wakeford, J. (2015). Developmental states and sustainability transitions: Prospects of a just transition in South Africa. *Journal of Environmental Policy and Planning*. doi: 10.1080/1523908X.2015.11077.
United Nations (UN). (2012). *Report of the United Nations Conference on Sustainable Development*. Rio de Janeiro, Brazil, 20–22 June 2012. New York: United Nations.
United Nations Environment Programme (UNEP). (2011). Towards a green economy: Pathways to sustainable development and poverty eradication. A synthesis for policy makers. Retrieved from www.unep.org/greeneconomy
United Nations Research Institute for Social Development (UNRISD). (2012). *Social dimensions of green economy*. UNRISD Research and Policy Brief 12. Geneva: UNRISD.
United Nations Research Institute for Social Development (UNRISD). (2016). *Policy innovations for transformative change: Summary*. Geneva: UNRISD.
Vogel, C., & Swilling, M. (2019). Climate change and vulnerability in South Africa: Sustainability transitions in a changing climate? In N. Mohamed *Sustainability transitions in South Africa*. London: Routledge.
World Bank. (2015). *South Africa economic update: Jobs and South Africa's changing demographics*. Washington, DC: World Bank.
World Economic Forum (WEF). (2019). *The global risks report 2019*. Geneva: WEF. Retrieved from http://www3.weforum.org/docs/WEF_Global_Risks_Report_2019.pdf
World Employment and Social Outlook (2018). Greening with jobs. Geneva: ILO.
Yale Centre for Environment and Policy. (2016). *Environmental performance index*. New Haven, CT: Yale University.
Ziervogel, G., New, M., Archer van Garderen, E., Midgley, G., Taylor, A., Hamann, R., Stuart-Hill, S., Myers, J., & Warburton, M. (2014). Climate change and adaptation in Southern Africa. *Climate Change*, 5, 605–620.

3 Mining

A laminated, dialectic methodology for identifying not-yet-obvious green skills demand

Eureta Rosenberg

Introduction

In 2015 South Africa's Mining Qualifications Authority (MQA) called for a study to determine what it termed the essential and scarce 'green skills' in the mining value chain, and the associated career paths, supply pipelines and challenges.

In *Sustainability Transitions in South Africa* (Mohamed, 2019), Montmasson-Claire noted that some industries are more likely to make sustainability transitions than others. For the researchers who undertook the MQA study (published by Rosenberg, 2015) a signal that the industry was interested in green skills was an opportunity to explore the potential for the kind of change heralded in Chapter 1 of this volume, through skills development. In South Africa mining had shaped the development trajectory of what was once the continent's richest economy, but it is now at a crisis point necessitating reinvention. As a sector expert at the Wits Centre for Sustainability in Mining and Industry noted:

> The future for South Africa's mining industry is uncertain. It is at a cross roads. If we get to grips with the current situation and put in place a robust plan of action, we can paint a picture of a thriving transformed industry, safer workplaces with more higher-skilled well-paid workers, more mechanization and automation, strong linkages upstream, sidestream and downstream to drive successful economic clusters, benefits equitably shared, local communities proud to host mining companies in their areas, regulations implemented transparently, and local and international investors attracted. … If we are going to bring about a more sustainable future … we are going to have to get everyone to think differently about our roles and responsibilities, and to reframe how we see the world and our priorities.
>
> (Digby, 2014)

The Green Skills for Mining study also presented an opportunity to pilot new methodology. It was informed by Lotz-Sisitka and Ramsarup (2017) and their review of International Labour Organisation (ILO) methodology guidelines and critical realist framings for research methodology (Bhaskar, 1993; cf. Chapter 13).

The methodology that emerged was further developed in subsequent demand studies (including those shared in Chapters 4 and 5).

The next section outlines the research process and data sources. Findings from and reflections on the contextual driver analysis are then shared. Next, findings from the value chain analysis are shared and discussed to highlight methodological processes and insights regarding the value of a laminated approach with a dialectical dimension. Finally, the chapter considers whether the study and methodology had the hoped-for transformative potential.

Research process and data sources

Figure 3.1 summarises the levels and stages of the research process. Note that while a sequence of stages was followed, the process was also recursive and iterative across the levels (i.e. we returned to earlier stages during and after subsequent

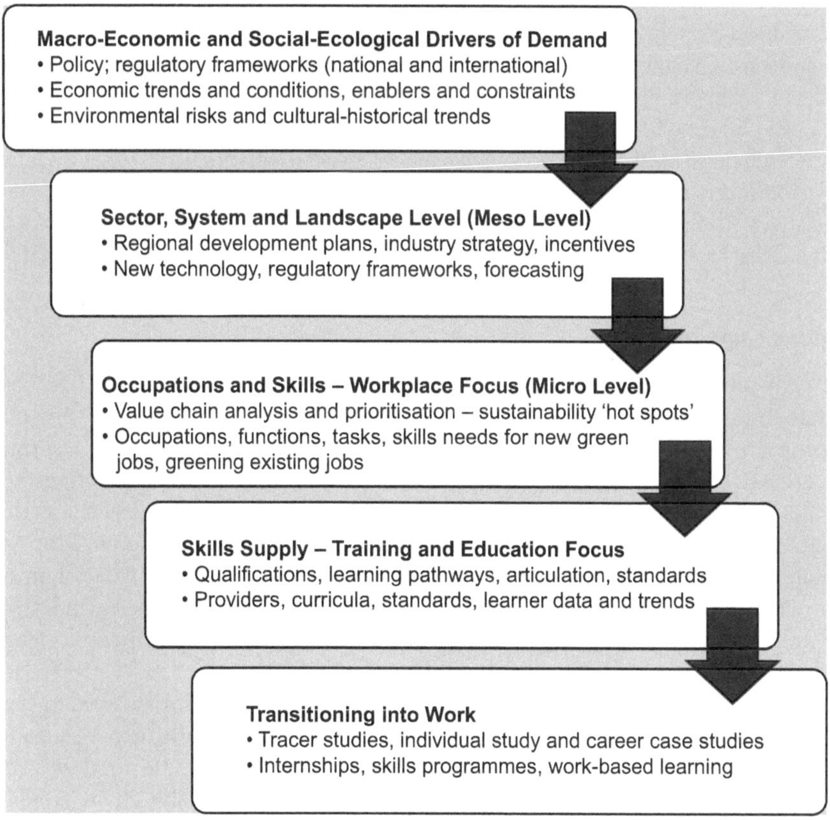

Figure 3.1 A multi-level, multi-method process for green skills demand determination
Source: Adapted from Rosenberg, 2015

stages, to revisit or collect further data in the light of findings of later stages, and analysed the data at the different levels, relationally).

An extended contextual driver analysis

In the first stage and at the macro level, the task was to identify macro contextual drivers affecting mining, and to consider their implications for green skills demand. This involved a desktop meta-review of analyses in company and sector reports and strategic plans, research papers, scenario studies and roadmaps. Interviews conducted in later stages provided further pointers.

To structure the contextual driver analysis, it was to some extent useful to apply the categorisation of the PESTLE analysis which South Africa's Sector Education and Training Authorities (SETAs) sometimes use for sector analyses, and which has been adapted over time from the 1967 text on analysing macro-environmental factors impacting business (Aguilar, 1967) which started with ETPS: economic, technological, political and social factors. The PESTLE refers to the Political, Economic, Social, Technological, Legal and Ecological drivers of a business strategy. We found it necessary to add History, as historical influences significantly shape current drivers in South Africa, resulting in another version of the acronym, PHESTLE.

We also noted that boundaries between social and economic, political and legal are often quite blurred, and policies can be discussed in relation to all other drivers. Furthermore, there was not a clear boundary between the macro and meso-level analyses (the next stage and level). Although Figure 3.1 suggests distinct stages, and these have proven useful in planning the study, in practice they overlapped.

Value chain analysis

At the meso (mid) level, the study focused on the coal industry itself. This analysis still involved a review of drivers of green skills, but started to focus through using an industry value chain analysis (Herr & Muzira, 2009) as the central method for identifying and mapping the green skills required. Constructing the value chain was based primarily on a more specific reading/analysis of the same data sources mentioned above for the implications for the coal industry. Mining companies themselves provide information about their commodity value chains, to which was added inputs from government, environmental agencies, industry think tanks and research institutes. The value chain analysis was further informed by a site visit to a coal mine and interviews (see below).

Also at the meso level (stage 2 in Figure 3.1) 'supply' data was gathered in the form of an overview of mining qualifications offered nationally, and aggregate level learner data. The 'supply' research dimensions of the study are not discussed here, but it is worth noting that this was done concurrent to value chain exploration, with iterative feedbacks. For example, as the value chain analysis showed rehabilitation as a priority area for green skills needs, we were guided to investigate the existence of courses and qualifications in mine rehabilitation.

Analysis of occupations and tasks

To better understand the skills needs identified at the meso level, micro level investigation was used to reveal details of the functions, processes and activities in the value chain. This was based on multiple primary and secondary data sources, such as interviews and email consultations with regulatory bodies, environmentalists, researchers and educators; a review of job descriptions in advertisements; documenting presentations on environmental functions in a coal company; scrutinising documents like the *Mining and Biodiversity Guidelines* (DEA, 2013), a report on the *State of Energy Research* (ASSAf, 2014) and the *South African Coal Road Map* (SACRMI, 2013); reviewing other studies including climate change research involving coal mine workers and the National Union of Mineworkers and studies on the impact of mining on surrounding communities; and undertaking workplace observations and interviews at a coal mine where we interviewed staff, several of whom had multiple roles:

- The Mine Manager
- The Project/Site Manager (also serving as Geologist and Environmental Manager)
- The Human Resource Manager (also serving as the Acting Safety Officer and Stores Manager as well as Skills Facilitator)
- Environmental Student (working as Stores Clerk)
- The Assistant Health and Safety Officer (also acting as Community Liaison)
- A Shift Boss with an additional role as Trainer.

This variety of sources, from different vantage points, proved important for a full and nuanced view of skills needs. This part of the study is presented by the last two stages in Figure 3.1, including individual study pathways and career stories.

Findings from the contextual driver analysis

Political and historical drivers

From the contextual driver analysis, a picture emerged of an old industry now experiencing deep ruptures, with local power struggles over the control of resources, and fossil fuels internationally increasingly framed as an anomaly. In Southern Africa, coal has been used since the Iron Age when early Africans burned charcoal to smelt metals. Large-scale commercial exploitation started in the mid-19th century. On South Africa's Highveld, where mines and power stations now dominate the landscape, African farmers thrived. When Europeans arrived, intimate knowledge of the land and advanced share-cropping systems gave the African farmers an initial advantage (Keegan, 1987). The advent of the first railways allowed them to sell their grain on the newly established mines, and expansion of the Cape colony created a demand for their wool. For them, the benefit of industrialisation was short-lived, however. European settlers started to outcompete with better access to resources including the tools of mechanisation. Over time, deliberate political-economic schemes locked Africans into a spiral of poverty that intensified with the introduction of the Land Act in 1913, and, in 1953, the educational strategy crafted by apartheid architects entrenched low skill levels and social inequality. From then on, livelihood options for black South Africans were to work on land – farms or mines – owned by others.

This history seemed relevant because the resultant low-skills, low-wage workforce remains a feature of the mineral economy into the present. Mining shaped South Africa's development path, bringing wealth and development on an impressive scale, but the benefits were unevenly distributed, and few Africans were able to work their way out of the disadvantage shaped by the associated complex of political, economic and educational drivers. After 1994, the socio-political landscape changed dramatically in some ways, and also stayed the same, as the minerals-energy complex remains locked into a path that can only benefit a few. This was further evident in the socio-political dynamics of the context.

Socio-political drivers

The African National Congress government introduced multiple policies to redress inequality. This included a Mining Charter aimed at the transformation of the industry, featuring targets for shifting ownership to black South Africans and 40% employment of black South Africans at management levels (DRM, 2004, revised in 2010). To address the historical uneven development that persists into the present, and, one could argue, to compensate for the disruptive nature and unequal benefits of mining operations, the Mining Charter also required companies to contribute to socio-economic development for host communities. Furthermore, the *Minerals and Petroleum Resources Act* (Act No. 28 of 2002, Republic of South Africa, 2002) requires companies intending to mine to submit a Social and Labour Plan for approval by the Department of Mineral Resources (DMR). This plan should include measures for local development and human resources development.

A Mining Charter Impact Assessment commissioned by the DMR in 2009 concluded that the industry was performing poorly on most elements of the Charter. For example, white employees continued to dominate top management and technical positions in mining, earning more than black South Africans. Failure to reach employment equity targets at management levels was attributed to skills shortages (Froneman, quoted in Peyper, 2013).

The social impact of the Mining Charter and other race-based equity transformation policies has been mostly in terms of empowerment 'deals' which failed to secure significant socio-economic development for the majority of black people working in and living near mines. With some significant exceptions, black people have not benefited from the wealth generated by mines, as highlighted by the deadly Marikana strike of 2012. Analysis of this tragedy pointed to a protracted ('locked-in', cf. Chapter 1) economic model that seemingly cannot afford to pay decent living wages to all its workers.

Both past and present governments protected mining because of its historic economic role. In 2015, when it was already in decline, the South African mining industry still contributed 7.7% to the Gross Domestic Product, around 15% to Foreign Direct Investment, 20% of private investment, 25% of exports and 1,4 million jobs (Baxter, 2016). This picture was changing, however.

Economic drivers

The mining industry globally experienced a metals price boom during 2000–2010. Since then, it has been under financial pressure. The boom has been followed by a reduced demand for commodities caused, for example, by China's economic trajectory moving away from commodity demand. A concomitant oversupply of several mined minerals has been driving lower prices (Baxter, 2016). The South African industry has been doubly impacted because its market share of global commodity production and reserves also declined, while input costs such as electricity increased, accessible seams were running out and infrastructure constraints were evident, along with declining productivity and demands for higher wages. Policy and regulatory uncertainty was also cited as a reason for the decline, along with what Baxter (2016) described as the "inappropriate application of regulatory tools" (see also Digby, 2014). Mining grew at a slower pace than the rest of the South African economy between 1994 and 2014. From 2012–2014 it shed 80 000 jobs, a fifth of its total workforce. The industry was loss-making in 2015 with a staggering aggregate net loss of ZAR37 billion.

Technological drivers

Economic considerations are driving investments in new technology, as are concerns about reduced productivity and demands for higher wages, motivating moves towards mechanisation and automation (Baxter, 2016). Resource depletion is a further driver (SACRMI, 2013). New tools, processes and skills are needed to mine increasingly inaccessible seams and to operate with water shortages becoming 'the new normal' in South Africa.

Internationally, the industry is also looking to mechanisation and automation to reduce environmental impacts and worker fatalities. In Chile, a lead copper producer, the Mining Council is looking at a large market share of mining technology, and to invest in new skills for mechanised mining (Araneda, 2017). The International Labour Organisation advised that "securing long-term employment in specific sectors like platinum relies heavily on the ability of South Africa to keep up with, and lead, global development of technology" (DHET, 2014, p. 132).

Another driver that will impact skills needs is linked to technology for the governance and administration of mining. The DMR introduced an online application and licensing system (South African Mineral Resources Administration or SAMRAD). In 2013–2014 this system was "in the process of being enhanced to ensure improved performance. A much more efficient integrated geographic information system (GIS) will be required to manage the balance between mining, agriculture, conservation and urban development" (DMR, 2014, p. 29). The need to address both social, economic and environmental concerns, is a key theme in the contextual analysis of mining. Identified environmental sustainability drivers are introduced next.

Environmental sustainability drivers

The South African National Environmental Management Act (NEMA) states that "the environment is held in public trust for the people, the beneficial use of environmental resources must serve the public interest and the environment must be protected as the people's common heritage" (Republic of South Africa, 1998).

Under the guidance of policies like NEMA, the Mining Charter features an environmental sustainability element, noting that *the social license to operate should include attention to the environment, along with workers' health and safety*. Auditors found that "maintaining a social license to operate" was the fourth most significant business risk for mining and is affected by poor environmental or safety performance (DEA et al., 2013).

The DMR's Mineral Regulation Branch grants mining rights and is responsible for legislation regarding mine rehabilitation (Eberhard, 2011). Environmental Impact Regulations published in terms of NEMA drive the need for license applicants to undertake either a Basic Assessment or a Scoping and Environmental Impact Assessment. In some cases, both the Minerals and Petroleum Resources Development Act (Act 28 of 2002) and NEMA require the identification, assessment and evaluation of impacts, and the determination of appropriate mitigation measures. All these should drive green skills demand.

South Africa's National Water Act (Act 36 of 1998) provides for an ecological and a social reserve, whereby government is committed to ensure that water is not extracted to such a point that the ecological functioning of the freshwater ecosystem is compromised, and downstream communities are guaranteed a minimum supply to meet basic water needs. Water use licences may not be issued if this will compromise these ecological and social imperatives. High levels of skills and competence are needed to inform and administer the associated law; however, a notable weak capacity in government to regulate its own policies was found to be a significant impediment to both development and environmental protection (NPC, 2012).

Mining and related activities have significant impacts on ecosystem services (e.g. delivery of potable water) and biodiversity; they often cause large scale and potentially irreversible habitat loss, including in strategic water resource areas. Deteriorating land and water quality in turn impacts other development options including agriculture and tourism, and human health.

Concerns over the country's emission levels are further significant drivers. The National Development Plan (NPC, 2012) refers to the transition required within the mineral-energy complex in order to direct South Africa on a clean development path. Coal-based energy makes up 80% of South Africa's greenhouse gas emissions which are very high compared to other coal-based economies at a comparative stage of their development. Industry partners (SACRMI, 2013) chose two drivers to predict future scenarios for coal: global climate change response and associated mitigation response. Climate change could affect mining through sanctions, trade tariffs or carbon taxes, reduced investment by sustainability-conscious investors, but also through physical resource constraints (like water).

The required reduction in coal consumption and therefore mining may shed jobs, or create jobs. The ILO supports the expectation that between 300 000–400 000 new jobs will be created in green industries in South Africa, noting coal mining as the only sector where decline in employment is expected (DHET, 2014). It advises that "Coal miners should therefore be targeted and empowered specifically for transitioning into greener jobs" (p. 132).

This advice is in keeping with the framing in the National Development Plan that "a shift to a green economy and more sustainable practices should not be seen in opposition to development, job creation and economic growth" (NPC, 2012, p. 181). The NDP argues that threats to the environment and the challenge of poverty alleviation are intertwined, and that a response to environmental issues should therefore not be framed as a choice between economic growth and mitigation of environmental impacts. The Plan is hopeful about a new energy architecture to ensure energy security and at the same time, a cleaner, safer environment, in which biodiversity is protected. Critical for skills planning was the statement that: "The transition to a low carbon economy depends on the country's ability to improve skills in the workforce" (ibid.). The National Development Plan also noted that "the transition to a low-carbon, resilient economy requires a capable state to lead, to enforce the regulation of greenhouse gas emissions, and to respond to the impacts of climate change" (ibid.). There is, however, a surprising absence of a coordinated strategy for growth. This is despite the existence of a South African Coal Road Map which predicted job losses under all envisaged future scenarios – to the order of *20 000–30 000 between 2010 and 2040*. While the Road Map hoped that these job losses would be compensated for by employment gains in a new development area, it also noted significant constraints to the planned development, including the lack of water. Thus industry experts and politicians called for new coal mining and processing methods that use less water.

While South Africa's Water Act and other environmental policies are well regarded internationally, policies developed in different sectors (e.g. energy, mining and environment) at times contradict each other, and may even have internal inconsistencies, or inconsistencies in application (DEA et al., 2013).

Methodological notes on the contextual driver analysis

Findings have been shared here in some detail to highlight methodological insights that also surfaced in other studies referred to in the book. It is worth noting the points listed below.

- The inclusion of history in the contextual analysis added depth to help understand present conditions and the seemingly intractable nature of socio-economic issues, even in the presence of new policy; depth analysis can indicate causality and emergence (cf. Chapter 13; see also the review of supply studies in Chapters 6, 7 and 10).
- A contextual driver analysis is not simply a summary of points from the literature; it requires analysis, including a consideration of the interactions between drivers, and *absences* (see below), for example, lack of implementation of strategies such as the Coal Mining Road Map, or gaps in the implementation of the Mining Charter, all of which point to high level skills needs.
- A *relational* analysis is needed, not only to identify which policies are in place and what they require, but to analyse how they relate to each other, whether they have internal contradictions, or contradictory implications, and what gaps remain.
- A contextual driver analysis would most likely not be completed at the start of a skills demand study, without being revisited in the light of later findings. That is, the relational approach calls for iterative analysis across the levels and stages of a demand study.

Discussion of findings from the value chain analysis

The need to extend the standard value chain

The coal value chain developed in the study (Figure 3.2) shows that South Africa's coal production supplies several local industries. The interconnections between coal mining and beneficiation, especially the minerals-energy complex, thus presents more of a 'value web' than a linear chain.

Figure 3.2 shows the coal value chain or web with the core mining functions starting with investment, research and development, exploration and mine planning, leading eventually to beneficiation in energy and other industries, extended to show parallel support functions such as training and reporting, and regulatory functions such as approvals and safety functions. Several 'values' such as local development and recycling were added to the standard industry value chain, not because they actually occur, but because, according to the sector policies, they *should* occur.

While this value chain in the past provided considerable employment, it is now at risk. It is estimated that coal production will stagnate in the next decade, become more expensive and, by 2020, will fail to fully meet the energy demand

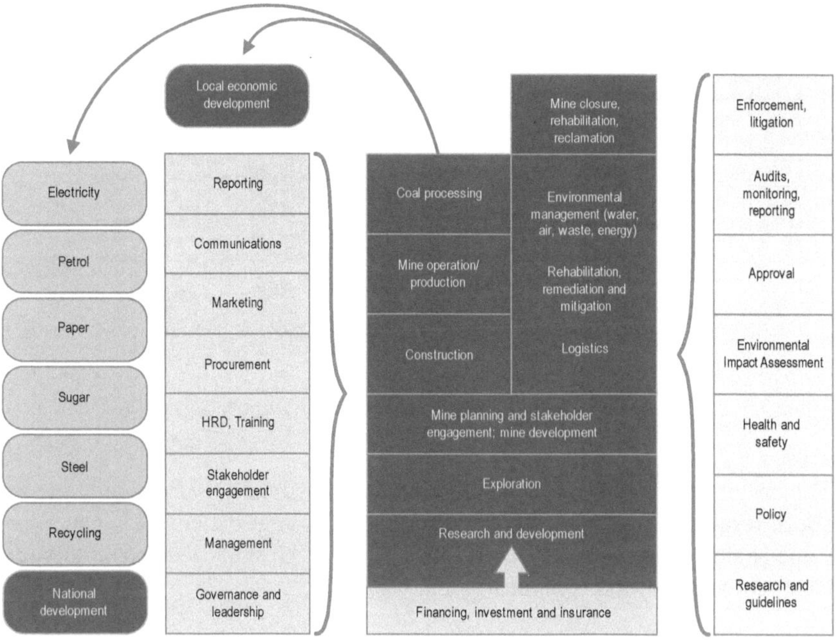

Figure 3.2 An extended value chain of coal mining
Source: Rosenberg, 2015

(SACRMI, 2013). Renewable energy production will become more competitive at the same time.

This value chain is also characterised by enormous waste and pollution. Much of the run-of-mine coal is not saleable. This amounts to more than 65 million tons of coal discards per annum; the South African countryside features some two billion tons of abandoned coal fines deposits (Kadzere, in Vermeulen, 2013). These dumps impact on water resources (acid mine drainage) and the surrounding environment, people and livestock.

The industry (SACRMI, 2013, p. 29) has therefore identified skills needs reflecting the need for innovation, such as:

- Know-how in thin seam mining (to unlock remaining reserves)
- Cost and water-efficient mining and beneficiation, particularly for multi-grade seams, including more research and development on dry beneficiation technologies
- Development and introduction of safer mining techniques
- Cost-effective and environmentally acceptable management of the volumes of discard and spoils
- Advancements in mine rehabilitation in order to better preserve biodiversity.

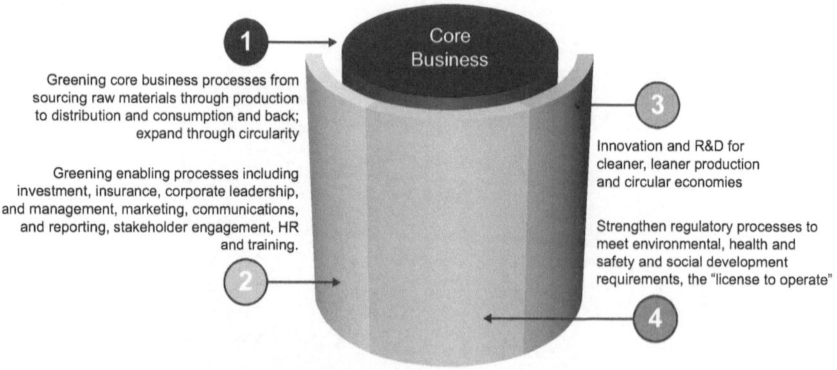

Figure 3.3 Extended value chains: Core business, enablers, innovation and regulation

The coal value chain analysis highlighted the importance of recognising innovation, regulation and other enabling functions as vital dimensions of industry value chains. This observation was made across green skills demand studies (cf. Chapters 5 and 6), and is summarised in Figure 3.3.

The value of micro level data

When the value chain analysis went into the detail of functions and occupations, it pointed to green skills being needed *everywhere* in the coal value chain. New green jobs and the greening of existing jobs are needed in head office for sustainability reporting; by engineers designing new technology and processes for mining, ore processing, waste treatment and waste management; by managers of mines and rehabilitation sites; by the regulators tasked with issuing licenses and enforcing compliance with health, safety, social and environmental guidelines.

Examples of associated occupations in the South African Organising Framework for Occupations (OFO) were:

- Regulatory skills for water use licensing include skills for demand determination such as: predictive modelling; catchment planning and management; and enforcement. Associated occupations: Water Use Officers; Environmental Scientists; Environmental Managers; Catchment Managers; Hydrologists, Geo-hydrologists; Hydro-geologists. These personnel may be employed by mining companies, consulting firms or government, in this case the Department of Water and Sanitation (notably, not the Department of Mineral Resources).
- Green procurement for corporate and mine supplies: Procurement and Supply Chain Managers; Sustainability Managers and Advisors; Sales and Stores Clerks

- Understanding broader environmental issues and safety and sustainability practices for mine workers and communities in a local context: Trainers; Health and Safety Managers and Practitioners; Labour Union Representatives; all workers
- Sustainable development planning and facilitation skills: Planners; Development Specialists and Social Scientists; Environmental Scientists, cross and transdisciplinary specialists, Community Engagement Officers / Process Facilitators; Trainers.

The micro level perspective provided an important nuanced understanding of the environmental management functions at the small mine visited during the study. Not only was considerable technical knowledge required across modalities (particularly for the single person tasked with environmental management across water, air and soil pollution as well as waste management) but also adaptive management skills. The Environmental Manager spoke of the need to make adjustments based on the outcomes of earlier actions; this was evident also in interviews with consultants playing similar roles for multiple mines.

Skills are needed at the mine, but also in the enforcement agency. The DMR's Annual Report (2013–2014) states in a section on 'Challenges faced by the Department':

> The need to enhance capacity to inspect and enforce compliance is one of the key challenges that is facing the Department. ... The high cost of training and the inability to retain staff (especially Mine Health and Safety Inspectors), has in the past few years resulted in a substantial loss of human capital investment.
>
> (DMR, 2014, p. 97)

The DMR has a shortage of Occupational Health and Safety (OHS) Inspectors, and finds itself unable to conduct all the necessary inspections because of this shortage (DMR, 2014).

One reason for the shortage is a growing number of mines encouraged by new ownership rules. Another, revealed in interviews, relates to salaries. OHS Inspectors must have a mining qualification (diploma or degree). Mining companies pay more for these qualifications than government does. When a government inspector arrives at a mine (where workers are chasing production targets), his/her work is often seen as an interference and the ensuing pressure on the inspectors is exacerbated by the fact that they earn less than their peers at the mine. For some there is little incentive to stay in a government job, if a corporate one is available.

It was important to have this micro level perspective on skills needs to understand the nuances and craft skills intervention recommendations. A bursary programme to address 'scarce skills' would not be successful if other factors driving scarcity, such as employment conditions, were not also addressed. Such insights are useful for a skills agency like the MQA (which in fact had supported the training of OHS Inspectors at postgraduate level for some time).

The need for a dialectic approach: Rehabilitation as priority intervention

This study identified as a 'hotspot' with great social and environmental transformation potential the rehabilitation of mined areas, both on active and worked-out mines. There is a 'lock-in' regarding the funding which is by law allocated for rehabilitation and mine closure, but 'traditionally' never actually applied. According to the *Financial Mail* (Seccombe, 2014), no closure certificate has been issued in 20 years. Mines are either sold on before closure or simply abandoned, evident in the estimated 6 000 derelict and ownerless mines on the South African landscape (Centre for Environmental Rights, interview data). Rehabilitation is seldom seen as more than 'mine closure', even though 'concurrent rehabilitation' is a regulatory requirement. This is not only a financial issue; mine closure operations are often disrupted as laid-off workers literally fight for the last remaining jobs (Seccombe, 2014). Incomplete or absent rehabilitation practices block the job creation and regenerative potential of more extensive and better managed rehabilitation and remediation efforts.

Should adequate funds be mobilised for this work, it would require a range of skilled people across occupations to work together to make a success of mine rehabilitation and realise its social and environmental benefits (see Rosenberg, 2015 for the associated skills needs).

Identifying rehabilitation as a significant place for intervention in order to meet vital social and ecological goals is an example of a dialectic process in a critical realist approach to demand analysis (explained in Chapter 13) in the sense that a social-ecological ill or 'absence' is being 'absented' (sensu Bhaskar; see Norrie, 2010).

The need for a relational approach to identifying scarcity

Analysis across the macro, meso and micro levels showed that skills requirements can be contested and that multiple perspectives need to be considered. The following illustrates this.

In addition to its core role of promoting mining, in 2013 the DMR also become the Competent Authority for regulating environmental applications and approvals and, as such, is responsible for reviewing mining applications, waste management applications, Environmental Impact Assessments and Environmental Management Plans. Mining companies submit applications through the South African Mineral Resources Administration (SAMRAD) online system mentioned earlier and applications are reviewed through a so-called One Environmental system.

In an emailed response to interview questions, the DMR maintained that it had adequate skills for these functions, with Master's level qualifications and upskilling through short courses. Another data source argued the contrary: that government lacked the necessary skills, partly because decisions about mining permits are made by relatively junior staff who are also under pressure to promote mining. A third data source, who had worked in the DMR before joining an environmental non-governmental organisation (NGO), elaborated on the systemic challenges of doing this work, for which, he believed, the recognised qualification does not adequately prepare one.

Concluding whether environmental compliance was a priority skill for the MQA to address, or even a 'scarce' skill, was a matter of iteratively analysing data from one level of the system (individual employee or micro level) in relation to a meso level of the system (DRM system for vetting approvals) and the macro level drivers (the dual mandate of the DMR and 'network' of other regulatory requirements) along with a history in which mining has had ascendancy over the Water and Environmental sectors.

Looking at the findings across the levels, one had to conclude that licensing approvals was a priority green skill. What would this skill entail? Tensions in an entity's mandate requires not only competence but 'wisdom' on the part of the staff, and top management support (and skill) to effectively execute the new function in an organisation (and wider sector with entrenched historical practices) required to make a fundamental shift (towards environmental sustainability). In addition to high levels of technical understanding it also requires social and relational skills in stakeholder management, strategic thinking and the ability to effectively call on resources and support (cf. also Chapter 9). Training interventions that build a community of practice among regulators, and link them with supportive other specialists, would probably be an important complement to sending individuals on short courses to enhance technical knowledge.

Concluding whether or not environmental compliance was a priority skill was thus a matter of *iteratively and relationally* analysing data from multiple levels of the system. This is what we mean by a *laminated* approach for demand determination. It differs from simply combining findings from multiple data sources, in the critical nature of the analysis, which also looks for 'absences' (cf. above and Chapter 13) and causal emergence – asking analytic questions about what has given rise to the contradictory findings, and what the underlying situation must be like for the contradictory empirical evidence to surface.

In conclusion: How transformative?

To what extent did the methodology employed in the study support transformative processes?

The study was intended to feed into the MQA's Sector Skills Plans and to support the MQA more broadly in thinking through the skills implications for mining at a potentially transformative juncture in the industry's development. We produced a report (Rosenberg, 2015) that took care to explain, not simply list, the findings and recommendations and also paid attention to the types of training that could address the types of skills needs identified.

Multiple data sources provided insight across levels in the system: From a macro level analysis of broad historical, social, policy and economic drivers, and a meso level focus on the coal industry and its value chain, to a micro level perspective on how environment-related work and training plays out in the context of a mine. While most green skills studies only focus on one or two of these levels (ILO, 2011), an analysis of the linkages and interplay between macro, meso and micro levels provided important depth and nuance for understanding the nature of environmental skills needs in mining.

The multi-layered perspective identified skills needs and scarce or priority skills needs that would otherwise not have surfaced, as they are 'not-yet-obvious'. The micro level perspective gave deeper understandings of skills needs including the *adaptive* nature of environmental management on the mine, and of causes of scarcity.

The study also demonstrated the need to extend the standard industry value chain and to look beyond skills needs identified by one set of employers, to adequately 'see' skills needs that may be hidden, for example, in innovation to treat or reduce waste, integrated social development and rehabilitation. By adding these, the study points to what (Bhaskar, 1993, p. 396) called "absenting the constraints on absenting the absences" (cf. Chapter 13).

Based on these insights, we made a submission in the Job Summit Colloquium convened in 2018 by NEDLAC (National Economic Development and Labour Council) on the need for learning to support a process to unlock the funding and job creation potential of mine rehabilitation. The NEDLAC Colloquium informed the 2018 Presidential Job Summit where mine rehabilitation was discussed. In the Framework Agreement arising from the Summit Report it has been agreed that "challenges with the financial provisions for mining rehabilitation will be resolved in an engagement between DMR, DEA and DWS" (Republic of South Africa, 2018, p. 43).

In that it raised questions that was unusual in many mining contexts, the study also had the potential to prompt new thinking and responses at organisational levels. The Human Resource Manager interviewed at the coal mine noted that it was probably necessary to employ more "environmentals". She did not have the language to describe the titles of their occupations, although her knowledge of the compliance requirements meant that she had a sense of the role they would have to play. The list of occupations provided in the research report (Rosenberg 2015) would support her.

Towards the end of the study, a water official who was consulted for the study, indicated that his department had decided to develop a course on water management for Mines. He wanted to know which university could develop such a course. We could guide him because the study provided a nuanced understanding of the challenges of working in a contradictory space faced by mine professionals, and pointed to those education providers that had developed educational methodologies that could help learners to navigate these challenges.

Whether or not more skilled professionals from inside the industry could help its transformation remains an open question. It does seem clear that coal mining has entered a phase that would be best addressed through radical transformation that will require considerable skills, including skills beyond the obvious.

Acknowledgements

Eureta Rosenberg led the research and write-up of the 2015 Green Skills for Mining study on behalf of Rhodes University Environmental Learning Research Centre. Heila Lotz-Sisitka (Rhodes University) and Presha Ramsarup (Wits University) made important conceptual contributions while Mucha Togo and Andani Mphinyane provided research assistance. The MQA funding and the support of Vuyokazi Mofu and Francis Lamola are also acknowledged.

References

Academy of Science of South Africa (ASSAf). (2014). *The state of energy research in South Africa*. Pretoria: ASSAf.

Aguilar, F.J. (1967). *Scanning the business environment*. New York: Macmillan.

Araneda, H. (2017). From natural resources to innovation: An opportunity for TVET in Chile. Fundacion Chile Presentation at the UNESCO TVET Conference, Tangshan, China, 4–6 July 2017.

Baxter, R. (2016). Mining in South Africa: The challenges and the opportunities. Presented by Roger Baxer, CEO of the Minerals Council of South Africa, to the Chamber of Mines of South Africa, September 2016. Retrieved from www.mineralscouncil.org

Bhaskar, R. (1993). *Dialectic: The pulse of freedom*. London: Verso.

Department of Environmental Affairs (DEA), Department of Mineral Resources, Chamber of Mines, South African Mining and Biodiversity Forum, & South African National Biodiversity Institute. (2013). *Mining and biodiversity guideline: Mainstreaming biodiversity into the mining sector*. Pretoria: DEA.

Department of Higher Education and Training (DHET). (2014). *National Skills Authority. National Skills Development Strategy III. Implementation Progress Report 2011–2013*. Pretoria: DHET.

Digby, C. (2014). Wits Centre for sustainability in mining and industry. *Mining Mirror*, October 2014.

Department of Mineral Resources (DMR). (2004, revised in 2010). *The Mining Charter*. Retrieved from www.dmr.gov.za/publications

Department of Mineral Resources (DMR). (2014). *DMR Annual Report 2013–2014*. Retrieved from www.gov.za

Eberhard, A. (2011). *The future of South African coal: Market, investment, and policy challenges*. Stanford, CA and San Francisco: Stanford University Program on Energy and Sustainable Development and Freeman Spogli Institute of International Studies.

Herr, M., & Muzira, T. (2009). *Value chain analysis for decent work*. Geneva: International Labour Office.

International Labour Office (ILO). (2011). *Skills and occupational needs in green building*. Retrieved from www.ilo.org

Keegan, T.J. (1987). *Rural transformations in industrializing South Africa: The Southern Highveld to 1914*. Basingstoke: Macmillan.

Lotz-Sisitka, H. & Ramsarup, P. (2017). SAQA *Bulletin: Learning pathways for sustainable development, the National Qualifications Framework (NQF), and lifelong learning in South Africa, 17*(1): 31–48.

Mohamed, N. (2018). *Sustainability transitions in South Africa*. London: Routledge.

National Planning Commission (NPC). (2012). *National Development Plan 2030*. Pretoria: NPC.

Norrie, A.W. (2010). *Dialectic and difference: Dialectical critical realism and the grounds of justice*. London and New York: Routledge.

Peyper, L. (2013). SA mining's next big crisis: BEE 2014. Miningmx, September 2013, p. 1. Retrieved from www.miningmx.com/page/special_reports/mining-yearbook

Republic of South Africa. (1998). National Environment Management Act, Act No. 107 of 1998. Government Gazette, Pretoria. Retrieved from www.environment.gov.za

Republic of South Africa. (2002). Mineral and Petroleum Resources Development Act, Act No. 28 of 2002. Government Gazette, Pretoria.

Republic of South Africa. (2018). Presidential Jobs Summit Framework Agreement. Retrieved from www.gov.za

Rosenberg, E. (2015). *Green skills for the mining sector*. Grahamstown; Environmental Learning Research Centre, Rhodes University.

Seccombe, A. (2014). Mine closure: Death and rebirth. *Financial Mail*, November 2014, 26–27.

South African Coal Road Map Initiative (SACRMI). (2013). The South African Coal Road Map. Retrieved from www.fossilfuel.co.za/initiatives

Vermeulen, A. (2013). South Africa's coal mining challenges. *Mining Weekly*, March 2013. Retrieved from www.miningweekly.com

4 Green skills for agriculture

A method for focusing demand analysis and prioritisation

Eureta Rosenberg

Context

My parents-in-law fulfilled a dream when they retired to Montagu, a small town in the Western Cape known for *veld*-flavoured lamb, fruit and wine. The mountains feature hot springs and footpaths named centuries ago by Khoi herders traversing the rugged terrain with their flocks on their way to water and grazing. My in-laws opened a shop. Khoi descendants, now farm and fruit factory workers, popped in to spend a bit of money. Then in 2012 a group of women farm workers called a strike to protest against the reduction of their wages. The strike soon spread down the Hex River Valley. Some farm owners dismissed strikers. Harvests had already been reduced by drought. Eventually my in-laws closed the shop. Montagu 'emptied out' – like rural areas in other times and places, adversely affected by environmental and economic conditions (Andrews, 2014; Misago, 2009).

One of the strikers pointed to the issues Khoi descendants face today which still revolve around water, land, economic well-being and the power to determine one's own path:

> It was an experience to be involved in the strike. When De Doorns came out on strike, we decided to join them in Montagu. … I told the police it is my right to protest. … It is my right to strike for a plate of food on my table. But, I want to say I am not here for R150 minimum wage. I am fighting for land and water because I realise that without land and water you have nothing and you will never fight your way out of poverty.
>
> (R. Plaatjes, quoted by Andrews, 2014)

Agriculture is entwined with South Africa's history of land appropriation and the inequalities that persist into current times. When Europeans arrived at the Cape in the 1600s, they encountered Khoi tribes with sheep and cattle. Further inland, Nguni people grazed their herds, and the Basotho practised sharecropping and produced wool. As news about the region's natural and mineral wealth spread, more Europeans arrived. Wars were fought; the most fertile and water rich land was captured. The 1913 Land Act confined Africans to peri-urban 'locations' and villages in the marginally productive 'tribal lands', or, in the case of the Basotho, to a barren mountain region. Without access to resources, Khoi, San and Bantu people were forced to labour on others' farms, in kitchens and mines. The same applied to the freed descendants of slaves brought from the East specifically to support farming.

It is an oft-told tale of a distant past. But the advent of democracy in 1994 has fallen short in giving all South Africans adequate access to water and land, named by the striker Plaatjes as the basic requirements for a pathway out of rural poverty. Only 7% of land has been 'redistributed'. The Land Reform Programme has had some notable successes, particularly where existing commercial farmers have teamed up with 'emerging' farmers. However, overall progress has been slow and according to one source (WWF, 2014) projects have shown a 90% failure rate.

Why is it so hard for new black farmers to start viable businesses? While there are multiple reasons (see, for example, Musvoto, Nahman, Nortje, De Wet and Mahumani, 2014), one is undoubtedly that farming in South Africa is a tough business. Many established farmers operate 'on the edge' of financial viability. This is related to the climatic conditions in this water stressed region, but economic models in which producers have limited control also play a role. Farmers carry the bulk of the input costs and risks, but receive the smallest portion of the profits of production. When a bunch of South African grapes is sold in the United Kingdom, 42% of the retail price goes to the retailer; 32% to the distributors; and 18% to the farmer. The profit margins are such that consolidating land into large units is an attractive option, as is mechanisation to reduce labour and extend the area under production. This is neither good for employment, nor for the protection of biodiversity in farmed areas (SmartAgri, 2016a).

While employment in commercial agriculture has declined in recent years, the sector is still a significant employer. Of the estimated 603 000 farm workers, most are employed in the Western Cape Province, which is suited to intensive farming. Much of South Africa's agricultural export production takes place here (Wesgro, 2014; Western Cape Government (WCG), 2014). The agricultural sector is also the largest employer in the region (WCG, 2014) and contributes to local economies, livelihoods and food security. But for the majority of the estimated 100 000 farm workers in the Western Cape employment is precarious. Many women in particular have only 'piece jobs' during harvesting and processing peaks. South Africans compete with migrants from Lesotho and climate stricken Mozambique and Zimbabwe. In the off-season all must survive on meagre savings and the goodwill of those with jobs. Substance abuse and violence are prevalent on these margins of economy and society (Misago, 2009).

To increase development and employment opportunities, the Western Cape government introduced a focus on 'climate smart agriculture' as one of the ways in which to both grow the economy and protect it from future risks. The region is highly vulnerable to the effects of climate change (WCG, 2014). Climate change predictions include a warming of 1.5–3°C by 2050, reduction of winter rainfall and an increase in variability of rainfall (Midgley, 2016). The Cape Town metropole was forced to apply extensive water consumption restrictions in 2017–2018 and farmers storing water for production were competing with urban users.

As part of its drive towards a green economy (Green Cape, 2016), the provincial government commissioned studies that resulted in the development of a Western Cape Agriculture Sector Climate Change Framework and Implementation Plan, shortened to the SmartAgri Plan (WCDEA, WCDA, WCDP, & SmartAgri, 2016). The green skills study drawn on here (Cobban and Visser, 2017) built on the findings of the Climate Smart studies, which interestingly did not include a skills dimension.

Overview of the study, scope and methodology

The Skills for Climate Smart Agriculture study (Cobban & Visser, 2017) explored the value chains and skills needs of three Agricultural sub-sectors in the Western Cape: pome fruit (apples and pears), winter rainfed grains (wheat and barley) and poultry (eggs and meat). While this chapter will focus only on findings regarding the poultry industry, the others are mentioned to demonstrate a need and process for progressively focusing the research gaze in green skills studies.

Taking its framing from the provincial government and its partners in the development of the SmartAgri Plan, the study initially started with the framing of climate smart agriculture. For conceptual coherence, the researchers had to consider a variety of framings for sustainable agriculture. These included conservation agriculture (with a focus on protecting biodiversity), organic farming and permaculture (which reduce harmful chemical use among other innovations), agro-ecology (with a strong social justice integration) and more. Climate smart agriculture tends to focus on technical solutions for production increases within prevailing conditions. It addresses sustainable development through three main thrusts: "(i) sustainably increasing productivity and incomes; (ii) adapting and building resilience to climate change; and (iii) reducing or removing greenhouse gas emissions, where possible" (Cobban & Visser, 2017). It assumes that greater production is necessary in order to address food security in the context of growing populations, resource degradation and climate impacts. A different approach to transforming agriculture would be to focus on righting historic injustices and laying the foundation for a more equitable future, which could also include the issues of distribution of the available land and resources and greater attention to the distribution of produce, while still taking climate risks, resilience and adaptation into account. While climate smart agriculture has a greater focus on productivity and scarcity, the green skills study opened up this framing by taking justice matters into account, for example, by considering the particular needs of farm workers, the new farmers in land reform programs and youth (Cobban & Visser, 2017).

Table 4.1 The research process for the Skills for Climate Smart Agriculture study

Level and scope	*Methods and focus*	*Data sources*
Macro level (agriculture sector in the Western Cape region)	Identification, analysis and comparison of contextual drivers, using document analysis	Predominantly grey literature from industry or commodity associations and government reports, and to a lesser extent academic literature. For example, the Western Cape Department of Agriculture, Forestry and Fisheries (DAFF) value chain reports for a number of agricultural commodities.
Meso level (three sub-sectors in the region)	Deeper desktop analysis of contextual drivers as well as key informant interviews, to identify leverage points ('hotspots') in the three commodity value chains	As above, in particular SmartAgri reports, commodity-specific documents, industry news articles, national documents such as the National Development Plan (NPC, 2012), value chain reports complemented by interviews with key informants.
Micro level (two workplaces)	Observations during site visits and further key informant interviews in the poultry sub-sector	Site visits including interviews on daily tasks and training, with managers and all employees on the workplace site.

Source: Adapted from Cobban and Visser, 2017

The overarching research questions were: Where is the biggest potential for greening in the agricultural commodity value chain, and what are the skills needed to unlock this potential? To answer these questions, the researchers followed the multi-level methodology developed in the broader Green Skills Programme (cf. Chapters 3, 5 and 13). The research process is summarised in Table 4.1.

Methodological detail

How were 'hotspots' determined?

The researchers started with an analysis of macro and meso level drivers and trends within the overall agricultural sector in the Western Cape. Faced with a multitude of significant drivers across such a broad scope, they realised they needed to focus on a sub-sector. They then narrowed their focus to the value chains of three commodities: poultry (broilers and eggs); winter grain; and pome fruit. These commodities were selected through a comparative review of commodities produced in the province, using the categories in Box 4.1.

Box 4.1 Criteria for comparing and selecting agricultural commodities as priority leverage points

- Climate change risk profile (the extent to which the commodity could be affected by projected climate change)
- Environmental impact (any negative impact of production on the physical environment)
- Economic value, growth potential and contribution to overall agricultural sector
- Employment potential
- Social impact (the positive or negative impact on local and vulnerable people).

Source: Adapted from Cobban and Visser, 2017

The chosen commodities also represented a range of farming practices (dryland, irrigated and livestock). In this early phase of the study, a desire to be comprehensive (*for scientific purposes*) and to catalogue *all* possible green skills needs influenced sampling decisions. It soon became evident, however, as it did in other green skills studies (cf. Chapters 3 and 5), that the number of green skills needs in extended industry value chains is very large. Green skills studies were undertaken in part to produce practical knowledge. That is, in addition to developing a replicable methodology, they had to inform decision-making among providers, skills planners, investors and government. For this praxiological purpose, it was important to *first extend the value chain and study its details on the micro level*, which meant a close qualitative look that would not be possible across multiple subsectors. Next, it was necessary to *prioritise* among the identified skills ('narrow down'). This process of prioritisation needed to be rigorous and justified, because it could influence policy or investment decisions. The progressive narrowing and prioritisation method used in the agricultural study is illustrated in this chapter.

Following a desktop based contextual analysis of macro level political, historical, economic, social, technological, environmental and legal drivers (cf. Chapter 3), the researchers worked methodically through each commodity value chain, identifying (using again the categories in Box 4.1) those stages of production that appeared to have potential for impacting positive change. These 'hotspots' or leverage points in the system (Meadows, 1999) were identified for further investigation to help understand the skill implications for people working in the occupations associated with that leverage point.

As in the case of the other green skills demand studies (cf. Chapters 3 and 5), the central approach to identifying areas of opportunity was the mapping and analysis of a value chain (Herr & Muzira, 2009), from inputs through to consumption and waste, and including processing and support services, regulation and technological innovation. In addition, the standard linear value chain was extended with the concept of a circular economy.

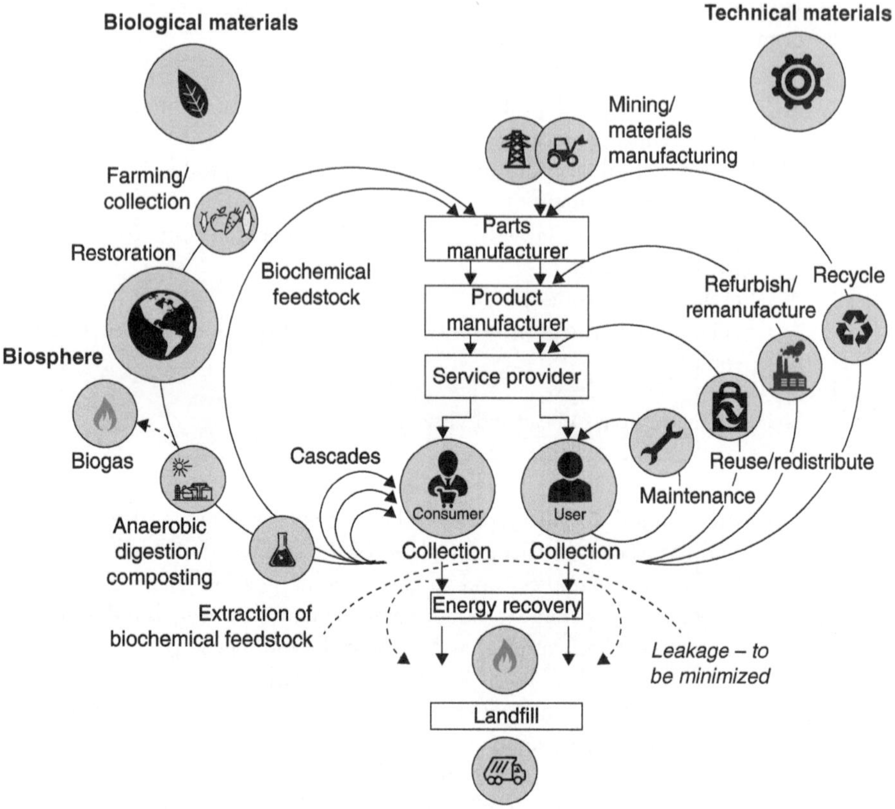

Figure 4.1 General features of a circular economy
Source: Ellen MacArthur Foundation, 2019

The concept of circular (or regenerative and restorative) economies (cf. Chapter 1; Figure 4.1) is helpful in identifying leverage points towards greater sustainability – opportunities for reducing wastage and creating additional value. Underlying principles (Ellen MacArthur Foundation, 2015) include waste being 'designed out', using energy from renewable sources, prices that reflect the full cost of externalities and diversity as a driver of versatility and resilience. A product, once used, is 'circled' back into the value chain to be re-used, re-furbished or recycled, or to have the biological components extracted and processed.

A life cycle analysis of a product or commodity is a related method, linking value chains and circular economies (e.g. Mulvaney, 2014).

The next step was to develop a nuanced understanding of the work tasks and skills involved. To this end, further data collection was undertaken at the identified leverage points through site visits and interviews with agricultural researchers, farmers and managers. Finally, the database of the national Organising Framework of Occupations (OFO) (cf. Chapter 12) was searched. Key occupations in the hotspots were assessed in more detail, and recommendations were developed for how the occupational description or associated occupational tasks could be adjusted to reflect greener practice.

Multi-criterion prioritisation among skills needs

This section further demonstrates the multi-layered methodology by focusing on the findings and the analysis undertaken to identify leverage points.

Applying the multi-criterion analysis outlined above, and *paying attention to the interactions between factors*, poultry stood out as an industry with high potential for job creation and related positive social impacts, while also being vulnerable to climate change.

The poultry industry is the largest agricultural sub-sector in South Africa, contributing ~22% of its income in 2012 (South African Poultry Association [SAPA], 2012). It is important nutritionally, providing around 66% of the animal protein consumed (Roosendaal, n.d.; SAPA, 2014). There has been growth across southern Africa in the consumption of meat and eggs (SAPA, 2012; WWF, 2010), accompanied by an increase in investments by predominantly large South African firms throughout the region (Ncube, Roberts & Zengeni, 2016).

Climate change may become a further driver in increased poultry production. The poultry industry's methane and CO_2 emission levels are lower than that of beef and mutton production (SmartAgri, 2016b).

According to its 2012 reports (ibid.) the South African poultry industry is also an important contributor to job creation, employing around 10% of all workers in agriculture. It provides direct employment for over 56 218 people and indirect employment to some 108 000 people through, for example, distribution. The industry is furthermore of interest in the context of just transitions and vulnerability to climate risks, because over 80% of producers are small, medium or micro enterprises. Some 2 300 small-scale farming enterprises have reportedly been established by government as part of its redress programme. *However, many of these projects have not been sustained*. One obvious concern is that the smaller new entrants struggle to compete with the large producers, who are already operating on a sub-continental scale.

A predicted increase in production means that it is especially important for the poultry industry to find ways to mitigate its environmental impact. The study identified that there are high levels of inefficiency and waste in the industry value chains, not only at the industrial-scale bird rearing and egg farms, but also at the abattoirs which discard animal parts. Electricity and water use is high, both to kill the animals and to clean up. Waste management and energy are significant input costs. For example, spent laying hens as well as male chicks are routinely treated as waste, because it is not profitable to transport them to abattoirs. Distances to slaughterhouses emerged as a significant factor mitigating against more widely distributed and small-scale poultry farming.

The large abattoir itself is a site of stress for animals and humans, featuring high staff turnover. The business model of most abattoirs also favours large consignments.

With these multiple factors in mind, the researchers identified small-scale poultry farming as a sub-sector for achieving social and environmental benefits through skills development. Within the sub-sector they then identified a number of leverage points with potential for simultaneously creating growth or developing new opportunities for start-up and marginal farmers, and stimulating more sustainable and climate-smart practices within the industry. Three of these leverage points, or 'hotspots', are listed below. Note that they are *linked.*

1. Creating circular economies using 'waste' as input to produce poultry feed, which would:
   - Reduce pollution
   - Reduce input costs
   - Make farming more profitable for small-scale new entrants.
2. Establishing local or mobile abattoirs, which would in turn:
   - Enable more widespread, small-scale and profitable production
   - Reduce waste by making the rearing of male chicks and sale of spent hens profitable.
3. Introduce environmental and ethical practices at small and large abattoirs to:
   - Reduce cruelty
   - Reduce waste
   - Reduce water and electricity usage.

Creating circular economies using waste to produce feed and fertiliser

The most costly farm inputs are animal feed, fuel and fertilisers. Retail prices of these are linked primarily to the oil price and the South African rand/dollar exchange rate, both of which are out of the farmer's control. This can lead to unanticipated price hikes in farm inputs. On an industrial scale, chickens are fed maize and fishmeal products. Both are costly and have negative environmental impacts. However, alternatives are possible that have less environmental impact, and reduce waste, by forming a circular economy (see Box 4.2).

Box 4.2 AgriProtein – example of a local circular economy

AgriProtein is a business converting food waste from restaurants and retailers to insect protein in the form of fly larvae as a natural poultry feed. On an industrial scale, the company produces dry larvae cakes as "the world's most sustainable and natural animal protein". On the *Farmer Angus Blog* site, a farmer interviewed for this study wrote in 2016:

"*A month ago we started feeding them [hens]… dried maggots from the geniuses at Agriprotein. These guys are taking 100 tons of Cape Town's daily food waste and turning it into nutritious fly larvae*".

This concept can be replicated on a local scale, where producers (or specialists among them) in township or village economies could use food and poultry waste to produce poultry food, at a fraction of the price of maize or green pasture, with less water. If food can become cheaper (breeding insects on waste as feed), rearing male birds as food may also become more viable than killing them at birth. The economy can be dispersed, with hygienically dried larvae sold to backyard chicken keepers. Economic viability will also be enhanced if the dumping of protein waste and other discards from poultry farming, which has a negative impact on water quality, is regulated and financially penalised.

If people are paid to remove chicken manure, it can become part of a circular economy too. It is already used on a small scale as garden fertiliser but with some attention to the value chain it could become a bigger asset in local circular economies. In 2008 the global demand for fertiliser surpassed supply. This, combined with the escalating price of raw materials, a rising oil price and shipping costs, resulted in record prices for fertilisers. A weakened rand further increased the prices of imported fertilisers. Local fertiliser production is influenced by international fluctuations and is under the control of a few large corporate companies (WWF, 2010). Farm-produced organic fertilisers and efforts to improve soil fertility have the potential to reduce input costs and the susceptibility of farmers to price fluctuations (WWF, 2010). This economic opportunity to reduce an expensive input could be enabled by the development of decentralised production, for example, fertiliser cooperatives as a value-add around small-scale poultry farming.

Establishing mobile and/or more localised abattoirs

Sustainable *waste management* is critical in poultry value chains due to the potentially hazardous material produced through the production and slaughtering process. Through the value chain analysis and interviews with farmers and experts, the slaughterhouse or abattoir was identified as a "hotspot" as it holds significant threats to the environment and can impact the success of small or new producers. The fact that they have high staff turnover confirms that they are not favourable working environments, either.

To support small-scale or new producers, more widespread, smaller slaughterhouses, or even mobile abattoirs could be considered. This would enable more widespread production that would be profitable for both small-scale farmers, and abattoir owners, as Box 4.3 suggests. Smaller slaughterhouses can also help to reduce waste in the value chain by making the rearing of male chicks and sale of spent hens more profitable (along with using food waste to produce chicken feed).

In the egg value chain, the laying hen who is 'spent' is treated as 'waste' because abattoirs charge more to slaughter them than they can be sold for. Sometimes spent hens are bought from egg producers and sold on informally in townships for meat, and slaughtered at home. Mobile slaughterhouses may offer better facilities – greater ease of the consumers, better health regulations and waste management. Centralised slaughtering could allow for waste collection to produce fertiliser for food gardens. Such units would need small-scale water and electricity supplied at each site, which opens additional entrepreneurial opportunities.

The emergence of 'outdoor' farmers with *free range poultry* was another niche within the sector, which is not discussed here due to space constraints. They too would benefit from mobile abattoirs geared up to take in smaller (niche market) volumes. The study highlighted therefore (a) the possibility of circular economies, and (b) the need for economic innovations, in this case reorienting to value small-scale economic interventions because of their local social, economic and ecological benefits, whereas in a conventional market framing they would be regarded as insignificant. It is an example of Raworth's (2017) call for an extension of the understanding of economic activity beyond the conventional market, to include households and the commons (cf. Chapter 1).

Why and how were these points in the value chain identified as hotspots? The process of prioritising and settling first on poultry and then on the small-scale farming poultry value chain has been described above, with the factors identified in document analyses and interviews being compared to consider their relative importance. This comparison was qualitative rather than quantitative. In other studies (Green Public Procurement, see www.greenskills.co.za) researchers used a quantitative scoring system to prioritise among commodities. An important factor in the success of the agricultural analysis was the availability of several in-depth commodity and climate change studies and scenario modelling on which the researchers could draw. These inputs from various specialists increased the confidence with which researchers could make comparisons.

From the value chain analysis, a number of points in the system design, production processes, distribution and consumption to 'end of pipe' were then considered as leverage points. A leverage point is a place in a system (the system being the commodity value chain or value web) where a relatively small intervention would result in a significant change in the desired direction (Meadows, 1999).

The green skills methodology involves a multi-criterion analysis (using the criteria in Box 4.1) to determine leverage points for skills related interventions. Not only were social, environmental and economic criteria included, the interactions between factors were also considered. For example, a particular economic impact may benefit large-scale and established producers, while disadvantaging new entrants or small-scale producers.

Connected leverage points may be particularly important when there are 'lock-ins' in a system at the regime level (cf. Chapter 1). Access to affordable slaughtering services is one such blockage in the system that disadvantages start-ups. The mobile abattoir could unlock that access and reduce transport costs. Producing feed from 'waste' reduces both the cost of feed and waste. Cheaper production means that spent and male chickens may gain value as food, rather than waste. More localised and mobile abattoirs would mean that live animals get transported over shorter distances, and smaller abattoirs can promote an ethical approach to killing animals for food. Historical practices include a daily blessing by an imam or priest to sanctify and purify the workers and prepare them for ethical workplace practices.

Managing waste for local feed production (insect protein) means that less land and water is used to produce grain or pastures for livestock.

All round, people and the environment benefit from:

- Economic value – higher income, less input costs
- Stable, sustainable local entrepreneurship and employment creation
- Environmental benefit – reduced waste, fuel, transport, water
- Social benefits – more dignified work for more people
- More ethical relationships between people and the animals they use as food.

The researchers carefully considered the feasibility of these innovations through site visits and interviews. They interviewed an outdoor chicken farmer, and the owner of a small abattoir who has turned it into a viable business employing several women (Box 4.3).

These innovations support economic activity at a different scale. While backyard or village poultry farming may be regarded as insignificant from a large economic sector perspective, it could happen on a significant (southern African) scale, and it is not insignificant for people who would otherwise be eking out a living on the economic margins. Raworth (2017) noted the importance of the household, the commons (environment) and good governance in supporting and complementing market economies (cf. Chapter 1).

However, it is important to note that the criteria on which the multi-criterion analysis is based must be carefully agreed beforehand, and should be in keeping with the chosen framework. A shift in framework will lead to different criteria and different skills being identified as priorities.

Box 4.3 Vignette of a small-scale abattoir

Hopefield Abattoir is run by a family who also rear their own poultry. Fifteen years ago, the abattoir ran for one day a week. It has grown to run throughout the week, with a capacity for 850 birds per day. Owned by a woman from a designated Black community, the business recently received support from Casidra (the Cape Agency for Sustainable Integrated Development in Rural Areas, an implementing agency of the provincial government) for upgrades and extensions to the facilities.

This small abattoir provides employment: For the owner-manager and her daughter who does the accounts; for five people who work in the first/starter room (slaughtering, dipping, operating the plucking machine and doing manual fine plucking); for two people in the second room, one person cuts up the carcass and takes out the intestines and the other person checks and rinses; for two people in the third room, who perform quality control checks, pack and label; for one person in the last room where the birds are packed into the freezer (this employee also helps with loading and off-loading). Once a week the abattoir processes kosher meat and gets in 12–15 extra women because the kosher process requires colder water and, therefore, more manual plucking.

Associated skills, occupations and specialisms

Innovations like mobile abattoirs require skills, otherwise they could be health and environmental disasters. They require, for example, skilled Operators and Health Inspectors.

According to the OFO, the Operations or Abattoir Manager (OFO code 134915) needs intermediate skills to support strategic business management through efficient and effective abattoir operation, best practice in all areas with a particular focus on animal welfare, hygiene and health and safety as well as reducing the negative environmental impact. Furthermore, they must be able to develop a competent, multi skilled abattoir team.

Climate change risks mean that the occupation of Operations/Abattoir Manager also requires water harvesting, recycling and management skills; skills for managing local clean, renewable energy sources to power the plant and energy efficient operations. Training for such skills could be provided through the South African Poultry Association (SAPA) and the Sector Education and Training Authority for Agriculture (AgriSETA) provided that these agencies can access this expertise. Operating the new technology – hygienic mobile slaughterhouses which dispose of waste safely, including through cyclical economies – may need new skills, but these still fall within the occupational description of the Operational/Abattoir Manager in the OFO. That is, a 'green specialisation' needs to become available as a training and qualification option, although it need not change the generic occupational description which already references best practice and environmental management.

A similar finding was made regarding other priority occupations identified in the study. It was found that Extensionists/Extension Officers with an understanding of new climate smart technology and options were in high demand among commercial farmers, but scarce, because this expertise is not yet included in agricultural extension curricula to a significant extent. Extension support is also vital (but seldom available on a continuous basis) for small-scale or informal producers (Nkuna & Mason, 2016). These individuals need to be able to advise farmers on waste management and cost-saving water and energy saving practices. The analysis of the OFO description of the Extension occupation found that the OFO need not change its description.

Another identified occupation with a 'green specialisation' is the poultry farmer who farms 'with nature' to produce 'free range' or 'outdoor' eggs. These practices differ significantly from producers who operate an industrial poultry business, but in the OFO they have the same occupation with little distinction between them. This would suggest the need for an occupational differentiation process somewhere in the skills system in which the emerging environmental specialisations – among farmers, Extension Officers and Plant Managers – can be recognised and named, so that investments in skills planning can take it into account (cf. Chapter 5).

Conclusions

In the Skills for Climate Smart Agriculture study the researchers journeyed from a more technical climate smart framing to a broader 'social justice' framing, inclusive of climate considerations, for the transformation needed to address social-ecological priorities in South Africa's agricultural industry. They also journeyed from a natural science research approach aimed at achieving full coverage of the industry, with representative samples, to a more focused, layered study to understand green skills needs at the occupational level required for skills planning. This smaller scope afforded them depth and detail to identify priority needs at the level of individual occupations, and the ways in which these need to be better represented in the skills system in order to guide investment in green skills development. Without skills investments, large-scale planning to make economies climate smart will fail to achieve benefits for all.

This research journey also affords other skills researchers the opportunity to learn about methods for prioritisation, both in terms of choosing a research focus and of making differentiated recommendations.

Acknowledgements

The *Skills for Climate Smart Agriculture* study (Cobban & Visser, 2017) was undertaken by the University of Cape Town's African Climate Development Institute (ACDI) as part of the Green Skills Programme, in partnership with the Environmental Learning Research Centre (ELRC) at Rhodes University and the Centre for Researching Education and Labour at the University of Witwatersrand (Wits REAL). We acknowledge the lead researchers Leigh Cobban and Zoë Visser of the ACDI; the inputs of Mark New (ACDI) and Presha Ramsarup (Wits REAL); Eureta Rosenberg (Rhodes ELRC) and Heila Lotz-Sisitka (NRF/DST SARChI Chair in Global Change and Social Learning) and all the research assistants in Cobban and Visser's study. The Green Skills Programme was funded as part of the Green Fund (Department of Environmental Affairs) through the Development Bank of Southern Africa where we thank Olympus Manthata and Eugenia Masvikeni for their support.

References

Andrews, M. (2014). *Sleeping giant is stirring: Farmworkers in South Africa*. South Africa: Trust for Community Outreach and Education, Agrarian Reform for Food Sovereignty.

Cobban, L., & Visser, Z. (2017). *Green skills for climate-smart agriculture: A case study of poultry, winter grains and deciduous fruit value chains in the Western Cape*. Cape Town: ADCI (African Climate Development Initiative), University of Cape Town. Retrieved from www.greenskills.co.za

Ellen MacArthur Foundation. (2015). Towards a circular economy: Business rationale for an accelerated transition. Retrieved from https://equalizer.co.za/planters/min-till-precision-sl-planters

Green Cape. (2016). *Green Cape agriculture: Market intelligence report*. Retrieved from www.green-cape.co.za/assets/Uploads//GreenCape-Agriculture-MIR-2016.pdf

Herr, M., & Muzira, T. (2009). *Value chain analysis for decent work*. Geneva: International Labour Office.

Meadows, D.H. (1999). *Leverage points. Places to intervene in a system*. Hartland, WI: Sustainability Institute.

Midgley, S. (2016). *Climate change and the future for agriculture in the Western Cape*. ACDI/UCT & WCDoA: SmartAgri, Drought Dialogue, 23 June 2016.

Misago, J.P. (2009). *Violence, labour and the displacement of Zimbabweans in De Doorns, Western Cape*. *Migration Policy Brief 2, Forced migration studies*. Johannesburg: University of the Witwatersrand. Retrieved from http://migration.org.za.

Mulvaney, D. (2014). Are green jobs just jobs? Cadmium narratives in the life cycle of photovoltaics. *GeoForum*, *54*, 178–186.

Musvoto, C., Nahman, A., Nortje, K., de Wet, B., & Mahumani, B. (2014). *Agriculture and the green economy in South Africa: A CSIR analysis*. Retrieved from www.csir.co.za/nref

National Planning Commission (NPC). (2012). *National Development Plan 2030*. Pretoria: NPC.

Ncube, P., Roberts, S., & Zengeni, T. (2016). *Development of the animal feed to poultry value chain across Botswana, South Africa and Zimbabwe*. *Wider Working Paper: 2*. Helsinki: United Nations University World Institute for Development Economics Research (UNU-WIDER).

Nkuna, C., & Mason, C. (2016, 18 April). *SAPA site visit* [Site visit and face to face interview with Charlotte Nkuna and Chris Mason].

Raworth, K. (2017). *Doughnut economics: Seven ways to think like a 21st-century economist*. London: Random House Business Books.

Roosendaal, B. (n.d.). *Poultry nutrition in South Africa*. *Rainbow Farms PTY, Ltd. South Africa*. Retrieved from www.thepoultryfederation.com

SmartAgri. (2016a). *Case study #2: Conservation agriculture*. Retrieved from www.greenagri.org.za/assets/documents

SmartAgri. (2016b). *A status quo review of climate change and the agricultural sector of the Western Cape Province: Brief for the Intensive Livestock sector*. Retrieved from www.greenagri.org.za/assets/documents

South African Poultry Association (SAPA). (2012). *The South African poultry industry profile 2012*. Retrieved from www.sapoultry.co.za

South African Poultry Association (SAPA). (2014). *The South African poultry industry profile 2014*. Retrieved from www.sapoultry.co.za

Wesgro. (2014). *Overview of Western Cape*. Cape Town: Wesgro.

Western Cape Department of Environmental Affairs (WCDEA), Western Cape Department of Agriculture (WCDA), Western Cape Development Planning (WCD), & SmartAgri. (2016). *A status quo review of climate change and the agriculture sector of the Western Cape Province*. Cape Town: WCDEA.

Western Cape Government (WCG). (2014). *Western Cape climate change response strategy*. *Environmental affairs & development planning*. Retrieved from www.westerncape.gov.za/

World Wide Fund for Nature (WWF). (2010). *Agriculture: Facts & trends South Africa*. Retrieved from http://awsassets.wwf.org.za

World Wide Fund for Nature (WWF). (2014). *Living planet report 2014: Species and spaces, people and places*. Gland, Switzerland: WWF International.

5 Surface coatings

Occupational analysis and green skills

Presha Ramsarup

Introduction

The need for green skills is a common thread running through all South Africa's green economy plans, yet there has been no catalytic initiative to build the capacity of the skills development system to produce proactively and strategically green skills for the green economy. Research to date (see DEA, 2010) indicates that while there is a skills development focus and a greening focus in South Africa, there is little articulated alignment between the two. Hence, more substantive understanding of green skills demand and utilisation is necessary to grow a green economy in South Africa (ILO, 2011) and to facilitate development opportunities at sector and national levels.

Responding to a need for better skills intelligence in South Africa, as well as the drive towards greater sustainability, in 2014 South Africa's Chemical Industries Sector Education and Training Authority (CHIETA) commissioned the CHIETA Green Economy research (Padalkar, 2014). This outlined that employers and organisations in the chemical sector still needed to develop an understanding of green economy trends and opportunities and argued for the need to develop employer awareness around the green economy. It further emphasised that "the content of existing jobs being orientated towards the green economy is a driver leading to a rise in the need for change in job content and the need for re-skilling of present staff" (Padalkar, 2014, p. 83), while recognising that there remained a need to better understand what these skills are.

The 2014 CHIETA report (Padalkar, 2014) also emphasised that the process of identifying green skills needs to start at grassroots level in a firm but that there needs to be multi-disciplinary engagement across the sector to ensure coordinated skills development.

Therefore in 2016 CHIETA initiated another study to assist it and provide the skills authority with a methodology to determine more accurately green skills priorities at both a firm and sector level. In designing this study, the skills planning manager emphatically stated: "Please don't tell me all jobs need sustainability training, that's useless to me. I can't do skills planning with that". This forced us as researchers to plan for a much more nuanced view of the skills needed to green work, which necessitated more intensive occupational lenses. We thus endeavoured to explore how the green skills needs could be more coherently represented within occupations, which are the unit of analysis for skills authorities in South Africa (to ensure systemic resonance).

The study furthermore aimed to investigate how green occupations, and the greening of existing occupations, could be more comprehensively represented in the national occupational frameworks to support more coherent skills planning within the chemicals sector. This study was undertaken by a small team of researchers from the University of Witwatersrand and Rhodes University (Jenkin, Rosenberg, Ramsarup, & Molebatsi, 2016) and built on a methodology used in an earlier study on green skills for the mining industry (cf. Chapter 3).

An initial review of the scope and complexity of the chemicals industry indicated that it would be necessary to focus on a segment of the sector. Using a set of social, ecological and economic criteria (cf. Chapter 4), a selection of sub-sectors (corresponding to different CHIETA Chambers) helped to determine where best to conduct the green skills study. Surface coatings was identified as a sub-sector and within it, *paint* was selected as one of the more prominent manufacturing sectors within surface coatings, the others being varnishes, enamels, oils, greases, waxes, concrete, lacquers, powder coatings, metal coatings and fire-retardant formulations. There are a number of sub-sectors within paints with a core application purpose. Following an assessment of the surface coating sub-sector, it was agreed to focus on *decorative paints* as it is the largest segment of the South African paint industry. The questions addressed by the study discussed here focused broadly on where in the industry value chain green skills would be required and which occupations would be involved.

The study used the same multi-level methodology and approaches for contextual driver analysis, hotspot prioritisation and value chain analysis as in the mining study. These elements will not be elaborated here; in this chapter the central methodological focus explored is the work on jobs tasks, occupations and occupational networks and how green skill needs were linked to the occupational classification system.

Overview of the paint sector in South Africa

The industry is valued at R13 billion and employs approximately 15 000 people. There are between 280–300 paint manufacturers in South Africa. Of these, the ten largest and most prominent produce up to 70% of the country's paint. Five of these ten are internationally owned and control 80% of the market share (Jenkin et al., 2016) which paints the picture of an internationally driven industry and an industry controlled by large companies. There are many smaller scale manufacturers in the industry, with a specialist or regional focus. From a skills development perspective it is important to take this into account, particularly with regard to the following: the financial ability to pay for and access training, and how opportunities for 'greening' skills for occupation may differ in comparison to the larger more prominent companies.

Paint serves important functional and aesthetic purposes, but it also has the potential to cause both health and environmental impacts. Low level exposure to paint may irritate or burn the eyes, nose, throat and skin and cause reactions such as headaches, dizziness or nausea, with unused liquid paint being classified as 'hazardous' and requiring appropriate disposal. The presence of lead is a critical issue in the paint industry and poses significant risks to human and environmental health. The study found industrial wastewater from lead-based paint manufacturing to be a large source of sludge water release; wastewater is created, for example, when the vessels used for producing paint are washed out (Jenkin et al., 2016). This leads to the generation of a paint-wash water effluent, which cannot easily be disposed of in the public sewerage system. An important observation by Jenkin et al. (2016) was that the regulation of lead-based paint and the management and disposal of wastewater within the lead-based paint industry are inter-departmental concerns. The first key stakeholder in the South African government is the Department of Health (DoH). In addition to the DoH, the Department of Environmental Affairs (DEA) and the Department of Water Affairs (DWA) have governance responsibilities for the leaded-paint sector and the wastewater by-product. This alerted us to the complex regulatory and legislative space within this sector.

The *Engineering News* of 15 June 2018 reported the government's notice of the intention to amend the Hazardous Substances Act of 2009, which was designed to eliminate lead in paint. At the time of the study, the Hazardous Substances Act classified lead as a hazardous substance and limited its use in paint to 600 parts per million (ppm). The amendment included modifying this to 90 ppm, a significant reduction that will almost completely eradicate lead from paint (Bulbulia, 2018). The amendments also include all industrial paints, previously excluded. This represents a significant reduction of environmental and human health impacts in the Paints industry with potentially significant skills implications.

The study found that innovation to reduce environmental impact in the paint industry is often led either by legislation or as cost reduction. The larger companies are governed by mother companies with international head offices and tend to have more developed environmental regulatory systems. How the environmental knowledge and skills that have been developed in these larger organisations can be transferred to the medium to smaller-enterprises to drive a reduction in environmental impact across differently sized businesses is a key challenge in the paint sector, with its large number of micro-enterprises.

Brief overview of methodology

Multiple data methods and data sources were used to generate a comprehensive and nuanced picture of the South African paint industry. This included document analysis, site visits and interviews. Extended site visits were undertaken at two major paint manufacturers. During these visits researchers observed key work operations and conducted interviews with Safety, Health, Environment & Quality (SHEQ) managers and officers, human resources (HR), technical and research and development (R&D) personnel. Further interviews were conducted with three other paint manufacturers, the Green Building Council, the South African Paint Manufacturers' Association (SAPMA) the South African Paint Industry Training Association and other strategic stakeholders linked to the use of paint. These are all discussed in more detail below.

We completed a contextual driver analysis (cf. Chapter 3) which reviewed the political, socio-economic, environmental and technological landscape within which the South African paint industry operates. The analysis illuminated the drivers that could influence the demand for green skills. We subsequently mapped out the South African paint industry value chain, from mining raw materials and manufacturing through to application and end-of-life. The approach adopted built on the value chain analysis undertaken by Rosenberg et al.'s *Green Skills in the Mining Sector* (2015) described in Chapter 3.

Following the value chain analysis, environmental and social hotspots (cf. Chapter 4) were mapped against the value chain components. In hotspot analysis, hotspots are identified by virtue of their environmental, social and ethical impact profiles, which thus enabled the study to filter and distil information to identify and prioritise hotspots for further investigation or action. We used these hotspots to assist us to map 'who works in those hotspot areas' and the main job families. This work guided the fieldwork and field researchers going into factories focused on observing and interviewing those workers.

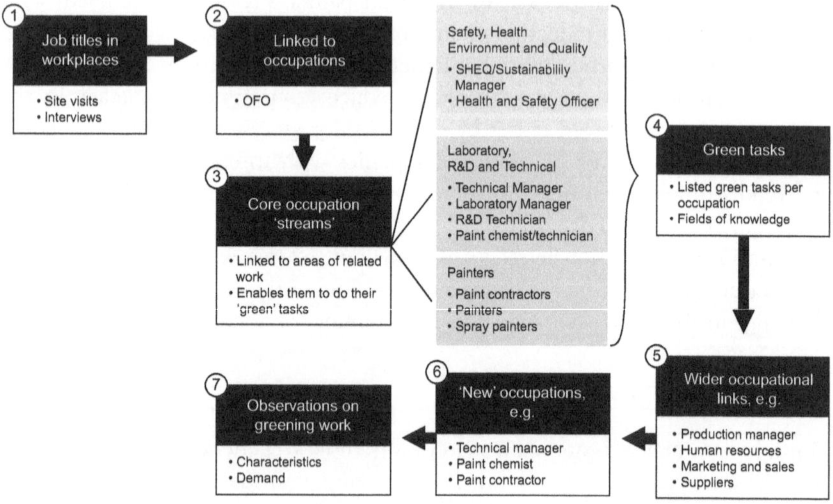

Figure 5.1 Surfacing green skill occupational insights

Discussion and findings

As described in Chapter 3, the demand studies discussed in this book rethought the traditional notion of a value chain to take into account social and environmental costs and benefits. We thus included support processes that run parallel to and in support of the core business of research and design; input; manufacturing; retail; and end use processes, such as: transportation (from raw materials to finished product); communications and reporting (including sustainability reporting); training; human resource development; and environmental management, health and safety management and regulation including enforcement. This also meant that the role players in the paint value chain include not only individuals in the industry, but also individuals in consulting firms, utility providers (water and energy as well as waste services) and (other) government departments: Health, Water and Sanitation and Environment (Jenkin et al., 2016). The report outlines that there is a critical need for both an external and internal role function of Environmental Health Officers (EHOs) (within and out of factory). The role of supporting EHOs through the DoH is to build capacity in Environmental Health Practitioners for conducting inspections or environmental health investigations and to ensure a reduction in conditions that constitute a health hazard (e.g. lead).

A further methodological step was the investigation of hotspots. The study helped to illustrate that there are significant environmental impacts at various stages in the paint value chain. This includes impacts during the extraction and processing of input materials and during manufacturing, which produces wastewater and sludge as noted; high energy use by the automotive paints industry; and health impacts on workers. During the end-use phase, there are also health impacts associated with applying paint and some components of paint (such as lead) remain hazardous long after application. Biocides and copper used in marine paints can affect the environment negatively during and after application. Unused or leftover paint that has to be disposed of is classified as hazardous waste. It is generally accepted that 10% of the environmental footprint of coatings is created during manufacture, with some 50% occurring upstream during the extraction and manufacturing of input materials, and 40% created downstream in application and disposal.

Building on the information from the value chain analysis and environmental hotspotting work, this section explores in further detail the occupational implications of greening work in the paint industry. The chapter will thus focus on how we synthesised findings on occupations and skills.

The review of occupations in the manufacture and application of paint draws heavily on interviews and site visits undertaken with industry stakeholders combined with analytical review of both the Occupational Information Network (ONET) (2016) and the South African occupation classification system in relation to work done on value chain analysis. Drawing on this desktop work, the interviews and site visits provided an excellent opportunity to understand and reflect on 'on the ground' occupational practices within organisations. We mapped the job families' (connected jobs) emergent from the document work and used this as a guide on the site visits to explore with whom people in these identified occupations worked and how the job tasks were related. Our quest was to determine, for example, with whom the environmental manager worked and thus who was connected to the environmental job family. Reilly (2004) proposed that 'job families' are groupings of jobs with similar characteristics, which focus on common competencies, skills and knowledge, or sometimes a similar purpose or process.

The discussion below highlights which occupations are core to environmental and social sustainability within paint organisations, and which different jobs consult and work together to enable social and environmental change in the implementation of their roles and functions. It also provides a detailed insight into the different green skills that are required for an occupational network.

In attempting to surface data on the job families, job titles from workplaces were collated and then mapped to relevant occupations on the Organising Framework Occupations (see Figure 5.1). Job titles were mapped, using the key tasks of the jobs, to relevant occupations on the occupational classification system. The data in this case reflects that in tracing the environmental tasks/work roles, the boundaries of job families were blurred and responsibilities tended to cross job families. Linkages were thus observed across traditional job families. In the study by Jenkin et al. (2016), these findings were presented as 'occupational networks'. Occupational networks reflect a group of occupations that work together to perform related environmental roles.

Skills needs were then analysed and presented to reflect needs across occupational roles in the network. It was not our intention to blur occupational boundaries but to reflect them as elements of multi-disciplinary practice that drive institutional collaboration.

The study identified three occupational networks, which are outlined below:

- **Safety, Health, Environment and Quality:** This network develops, implements and monitors an organisation's environmental, health, safety and quality strategies, audits, reports and environmental training. This will cover issues such as waste, water, energy, workforce and surrounding community health. Key occupations in this network include:
  - Safety, Health, Environment and Quality (SHEQ) Practitioner
  - Sustainability Manager
  - Environment Officer
  - Environmental Health Officers
  - Health and Safety Manager or Officer.
- **Technical and Laboratory (R&D and Technicians):** This network sets and manages product quality and technical and process technology standards for the manufacturing site. They transfer formulations to production and are involved in planning, organising and implementing activities and the research or production laboratory. The team develops innovative new solutions to environmental and health impacts associated with paint – including

water-based alternatives to solvents, lead-free paints or 'greener' products that can reduce the temperature of buildings. Key occupations in this network include:
    - Technical Manager
    - Laboratory Manager
    - Research and Development Manager
    - Research and Development Laboratory Technician
    - Paint Chemist.
- **Painters and Contractors:** The main purpose of this occupational network is to prepare a surface for painting, to apply the paint in a knowledgeable and skilled manner and to clean the equipment used. Contractors oversee a team of painters in or on larger operations, such as a building site. Key occupations in this network include:
    - Painter
    - Spray Painter
    - Vehicle Spray Painter
    - Paint Contractor.

Table 5.1 illustrates the Safety, Health, Environment and Quality network and the collaborative work tasks observed here. This observation highlights that the 'environmental worker/job' was in practice dispersed across a range of occupational roles having systematically integrated sustainability practices across and between their traditional occupational practices (Ramsarup, Rosenberg, Lotz-Sisitka, & Jenkin, 2018).

These observations reflect the emergence of shifts in workplace occupational practices and as Ramsarup et al. (2018, p. 159) noted, the observations on occupational networks further reflect "the ways in which the stock of knowledge in organisations and wider systems shifts". This has implications for training and how upskilling practices within workplaces are conceptualised.

Green jobs are not homogenous, hence they have different skills needs. Understanding these skills needs in relation to the greening of work is difficult without consistent, coherent and comprehensive occupational information. We therefore used a more nuanced typology to 'look at' green work; the typology was adapted from the International Labour Organisation (ILO, 2011) to align it with occupational work within the occupational classification framework in South Africa. The typology used to examine the scope of green work includes: 1) core green jobs which are specially defined as green jobs from a purpose, product and process point of view; 2) emerging specialisation in existing occupations where some specialisation in green skills is required to support sustainability oriented changes in the workplace, if not all the time; and 3) jobs not traditionally regarded as green but which have latent potential for supporting sustainability and sustainable value.

The second of these requires a framework to look more deeply into skills specialisation within an occupation. Thus, to inform the development of the occupational profiles in the paint study, Greenwood's (2008) articulation of four areas

Table 5.1 Key occupations and core green tasks for the Safety, Health, Environment and Quality network

Key occupations in this network	*Core green tasks for this network*
• SHEQ Practitioner • Sustainability Manager • Environment Officer • EHO • Health and Safety Manager or Officer	• Develop or execute strategies to address issues such as energy use, resource conservation, recycling, pollution reduction, waste elimination, transportation, education, and building design (ONET, 2016). • Drive lean management throughout the business, and raise awareness accordingly. • Application of, and compliance with, environmental, health and safety legislation and regulation e.g. ensuring workforce wear protective clothing and masks. • Undertake audits of key (Tier 1) suppliers in terms of their 'green' practices e.g. packaging suppliers. • Contribute to and/or co-ordinate reporting of sustainability-related activities (environmental and social) in e.g. annual corporate social responsibility reports) to internal and external organisational stakeholders and investors. • Research environmental sustainability issues, concerns, or stakeholder interests.

Source: Jenkin et al., 2016

used to review skills specialisation in an occupation was also used, namely: 1) the field of knowledge required; 2) the tools and machinery used; 3) the materials worked on or with; and 4) the kind of goods and services produced (ILO, 2011, p. 95).

Using these analytic lenses across the multiple levels of data, notably the skill need was not visible in skill profiles or workplace skills data or directly from employer perspectives. It required an iterative analysis drawing from the drivers as to what would be needed to improve regulation to envisage different manufacturing processes. This retroductive work (cf. Chapter 13) thus enabled us to determine the required skill intervention and the dimensions of occupational change to be considered with respect to greening occupations within the paint sector.

The discussion below reflects the fact that a large number of occupations are relevant to 'greening' the paint industry as reflected by Jenkin et al. (2016).

Decorative paint application

Existing skills and occupations are important

In the application (end use) of paint, there are a large number of occupations which require competencies that would make this part of the industry value chain 'greener'. For example, if painters are skilled in preparing diverse surfaces and applying paint correctly; if construction site managers and contractors are skilled in purchasing the correct paint in the correct volumes and in guiding and supervising painters correctly; if architects are skilled at prescribing appropriate paint, and if quantity surveyors can calculate the correct volumes to purchase, then applied paint would last longer and there will be less waste paint going into the hazardous waste stream.

New knowledge area for existing occupations

All the above professionals, technicians and artisans involved in the purchasing and application of paint, as well as marketing and retail staff, also need to know about and understand that there are more and less safe paints (e.g. lead-free paint and paint containing lead) and more and less eco-friendly paint (e.g. water-based paint) and to the correct way to work with the safer options. Innovations such as 'visualiser apps' also help consumers to waste less paint by reducing the number of inappropriate purchases.

Automotive paint application

Existing skills are important

The application of spray paint holds significant risks for workers, who need the necessary knowledge and skills to protect their health by using the correct equipment, including protective gear, in the appropriate manner. SHEQ staff at all levels need to be highly skilled at controlling environmental conditions like ventilation in the workplace, and health staff need to be skilled at picking up health issues early.

New knowledge area for existing occupations

Painting motor vehicles is currently a high energy usage activity. Given the risks of climate change and the rising cost of electricity, the automotive manufacturing industry is looking for ways to reduce electricity use. This involves both using existing equipment more efficiently and introducing new production processes. Engineering managers with particular specialisations in energy are needed to implement these changes, and employers report people with these skills as scarce. It is likely that new knowledge and competencies will need to be added to existing engineering occupations and plant operators.

Regulation of lead

Scarce skills in local government affecting the paint industry

SAPMA reports, confirmed by the literature, indicate that the use of lead in paint is poorly regulated in South Africa. While companies with international links are influenced by international legislation to eliminate lead from paint, other companies, in particular smaller ones, are less inclined to make this change, partly due to the higher cost of producing lead-free paint, and partly because the existing regulations (prohibiting the use of leaded paints in schools and playgrounds, for example) are not being enforced. Paint containing lead is sold at a lower price than the safer alternatives, which means unfair competition in the industry and a high risk of leaded paints being utilised in projects where cost is a particularly important consideration, for example low cost housing. This problem is exacerbated by poor knowledge of the need to use lead-free paint among contractors and other occupations among the end-users of paint. The *key scarce skill identified here, however, is the Environmental Health Officer* in local government. This officer needs to inspect and report on the presence of lead in paint in areas where it is deemed unsafe (along with eight other health related tasks). The Public Service Sector Education and Training Authority (PSETA) in its recent Sector Skills Plan reported 246 Environmental Health Officer vacancies. It is also acknowledged that it is challenging for existing Environmental Health Officers to effectively execute all their tasks. Regulations defining the Scope of the Profession of Environmental Health (DoH, 2013) outline the functions of Environmental Health Practitioners as including the Nine Municipal Health Functions that are required by District and Metropolitan Municipalities: Water quality monitoring; Food control; Waste management; Health surveillance of premises; Surveillance and prevention of communicable diseases, excluding immunisations; Vector control; Environmental pollution control; Disposal of the dead; and Chemical safety. This broad scope of tasks, combined with job vacancies, explains why the application of lead free paint is not well enforced.

It would be important to investigate the 'activity system' within which these officers conduct their work, to qualitatively determine the exact skills gaps and needs, and at what level(s) they occur, as well as the supply of Environmental Health Officers and the reasons for the vacancies. This occupation has an impact on the industry and its ability to operate fairly and within national and international safety regulations.

Manufacturing of paint

Existing skills are important

In addition to looking after human and ecosystem health, greening the economy is inter alia about making production processes cleaner, leaner and more efficient, reducing pollution and scaling down the input of resources. All these aims are addressed by existing industry practices such as reclaiming paint elements from dust or wash water to be re-used in making more paint. The skills of cleaner and leaner production, in the occupations of Paint Technician and Plant Manager should be retained and extended. Reducing energy and water use is similarly relevant here, as 'green' skills to be integrated within existing occupations.

New knowledge area (related to innovation) for existing occupations

The waste from paint manufacturing (washing of containers, discard of leftover sludge) presents a significant environmental challenge to companies, to the Department of Water and Sanitation that issues water use licenses and monitors compliance with water quality standards and to local governments whose landfill sites receive sludge and solid waste. The skills required in the water sector and the occupations involved were not investigated in this study. The occupation of Chemical Engineer, with a specialisation in waste management, remediation and design of new products, seems vitally important for the industry to be able to innovate, thus making its product safer for people and the environment.

In the occupational table (Table 5.2) we show how we attempted to link specific green skill needs identified in the workplace to the more widely framed occupational classification frameworks, which are significant to system change dynamics. Table 5.2 also illustrates how the study connected micro and macro and the critical need for data from multiple analytic levels.

The analytical work linked to occupations in this case example raised evidence of the following:

- Defined environmental job families were difficult to discern. Environmental responsibilities crossed job families. Environmental roles were characterised by institutional collaboration between different job roles. This collaboration meant boundaries between environmental job families were blurred.
- Environmental knowledge in the workplace was thus dispersed between different job roles and families.
- Underpinning skills and knowledge of greening require linked roles across related occupations (mentioned above, and also discussed further below).
- The integration of sustainability practices appeared to prompt elements of multi-disciplinary practice thus driving improved utilisation of skills in workplaces.
- Making decisions on appropriate skill interventions requires a nuanced idea of what skill specialisation is required – knowledge, products, process or tools.
- The majority of green skills interventions require greening of current occupations and the support of skills development that can foster more sustainable practices (what people do in jobs and their knowledge, skills and attributes). These need to be tackled within a systemic approach in the workplace and within the value chain (i.e. significant change is unlikely if only one person is implementing sustainability practices unsupported by the connected work roles).
- There are three focal areas that provide opportunities to enrich and increase workforce capability to support a more sustainable paint industry: risk, regulation and innovation. These need to be systemically linked across the value chain.
- The greening of specialist environmental roles was difficult to determine currently, and would probably need a more in-depth study, especially if all three categories of existing green jobs, greening of existing jobs and latent potential for greening work are to be adequately analysed (Ramsarup et al., 2018).

Table 5.2 Occupational tables depicting green skills

Occupation name: Sustainability Manager				
OFO *major group*	OFO *sub-major group (code)*		OFO *unit group (code)*	OFO *occupation code*
Managers	Administrative and Commercial Managers (121)		Human Resource Manager (1219)	121909
Segment of the value chain	*Skills band (high/intermediate/low)*		*Shade of green (dark/light/tint)*	
Manufacture	High		Dark	
Occupation description and purpose	Plans, organises, directs, controls and coordinates sustainability or environment-related risk assessments, develops methodologies to assess the viability or success of sustainability initiatives and the monitoring and evaluation thereof (DHET, 2015).			
Greening the occupation recommendations	Field of knowledge	- Needs to know, and keep up to date with, all necessary SA environmental, health, safety and quality assurance legislation and regulations. For a company with a parent company located internationally, this will often include a knowledge of international legislation and how to apply it (AkzoNobel, 2014; Axalta Plascon, 2016; Kansai Plascon, 2014, 2016b; Plascon, 2015). - Needs to follow and applyenvironmental management systems (EMS) in line with ISO 14001:2004 Environmental Management Systems Programme. - Needs to have ability to represent the organisation in negotiations and at conventions, seminars, public hearings and forums. - Needs to have ability and knowledge to collate information and progress from team for reporting purposes, and to develop programmes for improvement, e.g. lean management. - Needs to contribute, develop, understand and implement corporate environmental/sustainability strategy (for global, to interpret at a local, SA-level). - Needs to have knowledge to supervise employees, volunteers or champions working on sustainability projects. - Needs to understand all aspects of sustainability – environmental impacts (e.g. water, waste and energy) and social impacts (e.g. worker health and community impact) – and be able to relate to financials.		

		- Needs to work with marketing to develop supply chain and consumer messaging on, e.g. how to reduce paint waste or calculate the amount of paint required for a job. - Needs to work with procurement and R&D functions to raise awareness of toxicity to human health and the environment and seek alternatives (IPPC, 2015a). - Needs to understand suppliers and their materials (raw materials and packaging). - Needs to work with and procure (contract) external specialists to provide guidance on compliance and legal matters for e.g. water and waste or water and hazardous waste monitors.
	Tools and technology	- Develops methodologies to assess the viability or success of sustainability initiatives. - Monitors and evaluates effectiveness of sustainability programmes. - Develops, adopts and implements tools for improvement – lean manufacture, Health, Safety and Environmental (HSE) audits, etc. (Jotun, 2012). - Drives lean management programmes according to lean management principles (Axalta Plascon, 2016).
	Goods and services produced	- Coordinates, contributes to or produces company environmental or sustainability reports. - Ensures compliancy, e.g. ISO14001 environmental management standards. - Identifies, prepares and delivers training. - Sets targets and standards, e.g. for the reduction of carbon emissions and consumption of key environmental resources such as electricity, water and landfilled waste.
Specialisms or alternative titles	- SHEQ Director (Axalta Plascon, 2016) - Chief Sustainability Officer - Safety, Health and Environment Officer (121206) - Health, Safety and Environment (HSE) Manager (Naidu, 2016). *Note*: for the larger companies, such as Dulux and Kansai Plascon, there are often different levels of sustainability managers within one company, for example: 1. A manager who oversees a number of plants within a region, such as SA or sub-Saharan Africa. He/She is responsible for overseeing SHEQ and sustainability managers at individual plants within his/her region. Often this regional manager will report to a global manager. 2. Individual plant SHEQ or sustainability managers, who lead a team of environmental, health, safety and/or quality officers. These managers reports to the regional manager (1 above).	

Occupation name: Safety, Health, Environment and Quality (SHEQ) Practitioner occupation and associated green skills

(*continued*)

Table 5.2 (Cont.)

OFO major group	*OFO sub-major group (code)*	*OFO unit group (code)*	*OFO occupation code*
Professionals 2	Health Professionals 22	Environmental and Occupational Health and Hygiene 2263	226302
Segment of the value chain	*Skills band (high/intermediate/low)*	*Shade of green (dark/light/tint)*	
Manufacture	High	Dark	
Occupation description and purpose	Coordinate the worker in his/her department to ensure that legal requirements are adhered to (OSH Act) and maintain the factory and Site SHE management programmes and systems. (Needs to integrate Lead Regulations, 2001 from Health Act.)		
Greening the occupation recommendations	Field of knowledge	- SA environmental, health, safety and quality assurance legislation and regulations (emphasise lead regulations, e.g. exposure to airborne lead; assessment of potential exposure; air monitoring; control of exposure to lead; monitoring disposal and transport of lead) - Needs to understand all aspects of sustainability and its relationship to health and safety – environmental impacts (e.g. water, waste and energy) and social impacts (e.g. worker health and community impact) and relate to financials. - Works with procurement and R&D, marketing functions to raise awareness of toxicity to human health and the environment and seek alternatives (IPPC, 2015b).	
	Tools and technology	- Develops, adopts and implements tools for improvement – lean manufacture, Health, Safety and Environmental (HSE) audits, etc (Jotun, 2012).	
	Goods and services produced	- Coordinates, contributes to or produces company environmental or sustainability reports.	
Specialisms or alternative titles	- Safety, Health and Environment Officer (121206) - Health, Safety and Environment (HSE) Manager *Note*: for the larger companies, such as Dulux and Kansai Plascon, there are often different levels of sustainability managers within one company, for example: A manager who oversees a number of plants within a region, such as SA or sub-Saharan Africa. He/She is responsible for overseeing SHEQ and sustainability managers at individual plants within their regions. Often this regional manager will report to a global manager. Individual plant SHEQ or sustainability managers, who lead a team of environmental, health, safety and/or quality officers. These managers report to the regional manager (1 above).		

Facilitating change in work will involve the creation of transformative niches, which are spaces for change within workplaces and hence jobs. Occupations are inherently transient, not permanent, constructs. They exist within wider societies and suffer from internal tensions and tensions with other economic and social interests (Standing, 2010). The chapter thus argues that the constituting of occupations and work roles in workplaces as transformative spaces offers a conceptualisation that clarifies a tentative notion of what a transformative space could enable within the discourse of greening work.

The data here has helped to show that these transformative spaces have different social processes that can catalyse change, namely learning, a network and a clear vision and understanding of the required change. Occupations as levers of change can help to destabilise the existing work practice lock-ins and hence facilitate the transition to sustainability.

Conclusion

This chapter has provided insights into some of the key occupations that are critical for greening in the paint industry. The further associated green skills insights linked to each occupational network provided possibilities for skills planning interventions, which can be supported by skills authority and the paint industry broadly. In summary, the green skills needs across occupational studies reflect that the majority of green skills interventions require greening of current occupations and the support of skills development that can foster more sustainable practices (what people do in jobs and the knowledge, skills and attributes). There are three focal areas that provide opportunities to enrich and increase workforce capability to support a more sustainable paint industry: risk, regulation and innovation. These will be central to greening occupations.

Acknowledgements

This chapter draws on research conducted in the National Green Skills Programme, and this particular study was funded by CHIETA. The author acknowledges the work of the research project team, which included Nicola Jenkin, Presha Ramsarup, Eureta Rosenberg and Palesa Molebatsi.

References

AkzoNobel. (2014). *AkzoNobel sustainability report 2014*. Retrieved from http://report.akzonobel.com/2014/ar/servicepages/downloads/files/akzonobel_report14_entire.pdf

Axalta Plascon. (2016). Axalta Plascon site visit, Port Elizabeth: 18 February.

Bulbulia, T. (2018, 15 June). [Review of Proposed amendment to legislation set to eliminate lead in paint, by Z. Mavuso]. Engineering News. Retrieved from www.engineeringnews.co.za/article/proposed-amendment-to-legislation-set-to-eliminate-lead-in-paint-2018-06-15/rep_id:4136

Department of Environmental Affairs (DEA). (2010). *Environmental Sector Skills Plan for South Africa. A systems approach to human capacity development*. Pretoria: DEA.

Department of Higher Education & Training (DHET). (2015). *Complete OFO Version 2015*. Pretoria: DHET.

Greenwood, A.M. (2008). *Updating the International Standard Classification of Occupations (ISCO)*. Geneva: International Labour Organisation (ILO), Bureau of Statistics.

Intergovernmental Panel on Climate Change (IPPC). (2015a). *Global Alliance to Eliminate Lead in Paints (GAELP)*. Retrieved from www.ippic.org/advocacy/global_alliance_to_eliminate_lead_in_paints

Intergovernmental Panel on Climate Change (IPPC). (2015b). *Marine Coatings Technical Committee*. Retrieved from www.ippic.org/advocacy/marine_coatings_technical_committee

International Labour Organisation (ILO). (2011). *Skills and occupational needs in green building*. Retrieved from www.ilo.org/wcmsp5/groups/public/---ed_emp/---ifp_skills/documents/publication/wcms_166822.pdf

Jenkin, N., Rosenberg, E., Ramsarup, P., & Molebatsi, P. (2016). *Green skills in the South African surface coatings sector: A focus on paint*. Research report for Chemical Industries Education & Training Authority. Retrieved from www.greenskills.co.za/wp-content/uploads/2015/07/FINAL-Green-skills-in-the-paint-sector.pdf

Jotun. (2012). *Corporate responsibility report*. Retrieved from http://cdn.jotun.com/images/2012-jotun-corporate-responsibility-report_tcm29-1363.pdf

Kansai Plascon. (2014). *Kansai Plascon Africa Limited – 2014 sustainability report*. South Africa: Kansai Plascon.

Kansai Plascon. (2016a). Kansai Plason site visit: Krugersdorp: 3 March [Site visit and face to face interviews].

Kansai Plascon. (2016b). *Plascon Spaces Showroom*. Retrieved from www.plasconspaces showroom.co.za/public/home

Mash, T. (2015). Sustainability in the coatings industry: What lies beyond 'business as usual' improvements? PCI Magazine, 1 April. Retrieved from www.pcimag.com/articles/100363-sustainability-in-the-coatings-industry

Naidu, P. (2016). Green skills in the paint sector: Dulux. 26 February [Telephonic interview].

Occupation Information Network (ONET). (2016). *11-1011.03 – chief sustainability officers*. Retrieved from www.onetonline.org/link/summary/11-1011.03

Padalkar, A. (2014). Identification of future skills needs of the chemical industries sector with reference to the green economy. Johannesburg: CHIETA.

Plascon. (2015). *Corporate social investment*. Retrieved from www.plascon.co.za/about/csi.asp

Ramsarup, P., Rosenberg, E., Lotz-Sisitka, H., & Jenkin, N. (2018). Green skills: Transformative niches for greening work. In N. Mohamed (Ed.), *Sustainability transitions in South Africa*, pp. 145–165. London: Routledge.

Reilly, P. (2004). *Job families: An integrating approach to reward and development?* Brighton: Institute for Employment Studies.

Rosenberg, E. (2015). *Green skills for the mining sector*. Grahamstown, South Africa: Environmental Learning Research Centre (ELRC), Rhodes University.

Standing, G. (2010). *Work after globalization. Building occupational citizenship*. Cheltenham: Edward Elgar.

6 Learning pathways into environmental specialisations

A boundaryless careers perspective

Presha Ramsarup

Introduction

With this chapter the book moves from demand studies (cf. Chapters 3, 4 and 5) to a greater focus on supply and transitioning of green skills. The relative newness of the environmental sector and the rapidly changing nature of environmental knowledge place constraints on traditional supply-demand methodologies and often surface incomplete pictures of sectoral green skills needs. Our research has found that a focus on transitioning helps to surface wider systemic articulation concerns as it provides both structural and systemic insights.

The cross-sectoral nature of environmental skills development and the complexity of planning for cross-sectoral skills of this nature imply inter-disciplinary foundations for green skills development. Understanding the transitional processes and outcomes is significant in deconstructing the mechanisms that shape the crafting of a career in the environmental sector.

In framing a response for analysing learning pathways into environmental work, Ramsarup (2017) argued that reflecting on professionals' educational and occupational progression is necessary within environmental occupations, and thus different dimensions of transitions emerge:

- From education to work (e.g. from higher degree into a job)
- From work to education (e.g. from job into further study)
- Within education and training systems (e.g. from a higher education formal degree context into an occupational directed [short] course/s)
- From occupational and work life into wider social processes (e.g. from engineering practice to engaging with societal processes that provide normative assessments of these). (Ramsarup, 2017)

This necessitates moving beyond a focus on the individual to focusing equally on structures and contexts that influence careers. To transcend this dualism there is need for a methodological approach that can examine how the system more broadly impacts the worker and which transcends traditional firm/organisational boundaries.

The inter-sectoral, boundary crossing nature of environmental jobs: A focus on the environmental scientist

People who work in environment and sustainability fields can be found in all walks of life, in all sectors engaged in planning for, and finding solutions to, issues that threaten the sustainability of the planet and its people. These issues are environmental, social and economic in nature and hence the work life and tasks of environmental workers naturally cross multiple sectors.

As most occupations in the environmental sector are focused on the complex nature of environmental concerns, environmental occupations often demand inter-disciplinary and inter-sectoral engagement (cf. Chapters 3, 4, 5 and 9). What is not known is the way in which the nature of these occupations structure the way in which careers and learning pathways are constructed.

Environmental occupations typically do not have clearly defined institutionally or occupationally determined pathways. The environmental sector is characterised by professionals navigating their own non-linear pathway into an environmental job (Ramsarup, 2017). This has historical antecedents, most notably the recent emergence of such occupations on the national landscape and the lack of dedicated systems of education and training development for the environmental sector which, as noted in the South African Environmental Sector Skills Plan, is a 'cross-cutting' issue (Department of Environmental Affairs [DEA], 2010).

This chapter focuses on the environmental scientist, who today works in environmental consulting firms, conservation agencies, all levels of government (national, provincial and local), private companies like mining and engineering firms, research institutions and some non-governmental organisations (like the Worldwide Fund for Nature [WWF] and the Endangered Wildlife Trust). Driven by emerging environmental legislation (cf. Chapter 2), there is a well-documented need for environmental professionals who can administer the requirements of the new environmental laws (DEA, 2010; Human Sciences Research Council [HSRC], 2009).

Despite the emergence of this occupation, very little coherent information exists on the emergence of environmental sciences as a discipline in South Africa. This may be attributed to its 'newness' and the field may therefore lack the reflexivity of longer-term emergence.

The challenges experienced in South Africa resonate internationally. Reviewing environmental sciences in the United Kingdom, Smith (2011) raised some important points related to the wide varieties of specialisations which, he argued, ultimately result in a diversity of understanding, knowledge, skills among environmental science graduates which has in turn created "uncertainty in employers' minds around the characteristics of environmental science graduates" (p. 27). Thus, after 50 years, employer understanding in the UK around capabilities of environmental science graduates is still very poor compared to more traditional academic disciplines (ibid.). More positively though, this has enabled environmental scientists to work in a variety of contexts thus providing routes to a variety of specialisation options. The analysis suggests that environmental scientists are working in what can be characterised as 'boundaryless careers'.

The environmental scientist career stories used in this chapter help us to understand how non-linear transitions emerge, the complexity of these transitions and the need to attend to broader institutional arrangements within and across education and training, the labour market and the workplace.

Boundaryless careers

This section shares theoretical perspectives on careers that are formed as a result of inter-sectoral or inter-disciplinary influences. In the literature the closest description of the nature of environmental work was found in descriptions of boundaryless careers. The boundaryless nature of environmental careers is accentuated by the rapid proliferation of emerging sectors, cross-sector collaborations and innovative occupational communities (Tams & Marshall, 2011).

This section shares what can be learned methodologically from this, but also some of the limitations of this research, as interpreted from the vantage point of the environmental sector in South Africa. Boundaryless careers evolved out of a need to respond to the growing inadequacy of the constructs framed by stable organisational careers.

Arthur and Rousseau (1996) are the originators of the term 'boundaryless career' and they noted six aspects therein:

- Movement across the boundaries of separate employers
- Validation from outside the present employer
- Sustained by networks and information external to current employer
- A break with traditional organisational career boundaries
- Rejection of traditional career opportunities for personal or family reasons
- Perception of a boundaryless future regardless of structural constraints.

Fenwick (2006) discussed two further elements of boundaryless work which are also relevant. She explained that boundaryless work involves a sense of specialised expertise being developed and offered which would mean that the career is marked by the development of specialised portable skills, knowledge and abilities. The second element is job mobility across multiple employers, erasing conventional boundaries defining job, workplace and employer.

Gunz, Evans and Jalland (2000) argued that careers have not become boundaryless; rather career boundaries have become considerably more complex and multifaceted. But this is not the only critique levelled against this career discourse; it is largely regarded as a discourse that works only for highly skilled jobs as career movement for lower skilled workers is largely intra-organisational (Roper, Ganesh, & Inkson, 2010). There is furthermore a strong focus on the mobile individual who voluntarily builds careers across boundaries. It is also critiqued for presenting boundaryless careers as only positive, neglecting to acknowledge those who struggle with adapting and those with negative career outcomes (Eby, Butts, & Lockwood, 2003). The focus on agency and rational choice has been critiqued, and some would argue that the importance of structural restrictions and how they affect career behaviour is underestimated (Forrier, Sels, & Steynen, 2009).

Developing methodology for understanding 'boundaryless careers'

Notions of systemic/institutional pathways and individual pathways have emerged across the literature. This apparent dualism presents a challenge to pathway work broadly and within the environmental sector more specifically.

Despite extensive use of the learning pathways concept in policy, both locally and internationally, in countries such as the Netherlands, Norway, Germany, Australia, New Zealand, Scotland and Denmark (McKenzie, 2000; Raffe, 2003; Vaughan, 2003), it remains a metaphor that is imprecise. Raffe (2003) acknowledged that the concept is a very useful bridge between policy and research, despite the fact that learning pathways alone are not a recognised theory or a rigorous theoretical tool.

In this research, Ramsarup (2017) used a critical realist lens which assisted with transcending the dualism between the systemic and the individual pathway through its emergent properties in open systems ontology (Sayer, 2000). Critical realists argue that reality is both stratified (real, actual, empirical) and differentiated; they are able to envisage structures and mechanisms operating at non-actual, but nevertheless real levels (cf. Chapter 13).

The stratified ontology (Sayer, 2000) of critical realism enabled pathways to be presented as a laminated totality (Bhaskar, 1993), which allowed one to work with social phenomena like learning pathways by explaining them at different levels or scales and acknowledging that in open systems a multiplicity of mechanisms (conditions, agencies), emergent at different levels of reality, is always involved (Price, 2012).

Critical realism notes that both the generative mechanisms of the real and the events of the actual are not necessarily reflected in the experiences of the empirical (Lindley, 2014). Therefore, one cannot rely only on the empirical to reveal the causal effects of the research phenomenon.

In the empirical domain, it is possible to identify experiences of knowing subjects (e.g. the environmental engineers who reflectively share their biographied learning pathway and how it came to be constructed). In the actual domain, one can identify events and their experiences by subjects (e.g. how a particular course or degree programme shaped the learning pathway of the environmental scientist or environmental engineer). In the domain of the real, one can identify both experiences and events, and the underlying mechanisms that generate these events and make them available to experience (for example, how the existence of an academic stream in university education allowed for the taxonomist to study taxonomy and thus become a taxonomist; or how the policy of transformation or the change in power relations in society influenced the chance to obtain a management position).

It is the multiplicity of mechanisms (conditions, agencies), emergent at different layers of reality (Price, ibid.) that generates the complexity suggestive of open systems.

Critical realist research should aim for a rich variety of perspectives of the same object; hence the need for what Bhaskar and Danermark (2006) called "laminated explanations". Understanding the complex nature of environmental learning pathways involved conceptualising the pathway as several ontologically different and irreducible levels. Ramsarup (2017) conceived of it as "ontological levels that cannot be separated" (Nunez, 2014, p. 111). Bhaskar and Danermark (2006) argued that conceiving of social phenomena as laminations would help researchers contend with reductionism as no one ontological level should be privileged in research. Therefore, they recommended case specific disciplinary ensembles depicting laminated totalities. Nunez (2014) suggested that there is no a priori way in which reality produces complexity; thus the building of a laminated system does not follow a particular ordered approach; rather its formation is in direct relation to the explanatory concerns and the reality that determine them.

This underlabouring enabled us to explore a perspective that goes beyond the colloquial currency that the concept of learning pathways has gained in South African education policy, which has focused mainly on 'pathway engineering' which represents fairly "flat, actualist accounts" or a "purely positive account of reality" (Bhaskar, 1993, p. 400). The dialectical critical realist (cf. Chapter 13) perspective necessitates in-depth research and deliberation that enables us to look beneath what we see (the way a learning pathway is or is not shaped).

In attempting to emphasise the need to examine different dimensions of a phenomena, the study conceptualised the learning work transitions in their laminated totality (Bhaskar, 2010, pp. 9–10). This is depicted in Figure 6.1 which illustrates how the transitioning idea has been built across the levels to enable the development of a deeper understanding of the transitioning processes and outcomes within environmental learning and work that shape environmental learning pathways.

In a study which probed this methodological absence in work and learning research Ramsarup (2017) focused on two cases (environmental engineer and environmental scientist). She studied how environmental engineers and environmental scientists gain access to core environmental specialisms. The cases were selected around two scarce skill occupations identified in the sector, as there was little understanding of how the scarce skill problem had arisen.

The research was designed in a multi-layered manner which, though presented here in a linear way, was much more iterative, interconnected and overlapping in practice (cf. also Chapter 3).

Figure 6.2 provides an orientation to the different levels of elements studied to make sense of the learning pathway of environmental engineers and environmental scientists. This emphasises the multi-layered, *relational and multifaceted nature of the environmental engineer's learning pathway* uncovered in the study. The 'multi-layered' analysis of the career stories is extended by *system analysis* (for example, analysis of the provisioning system of engineering qualifications;

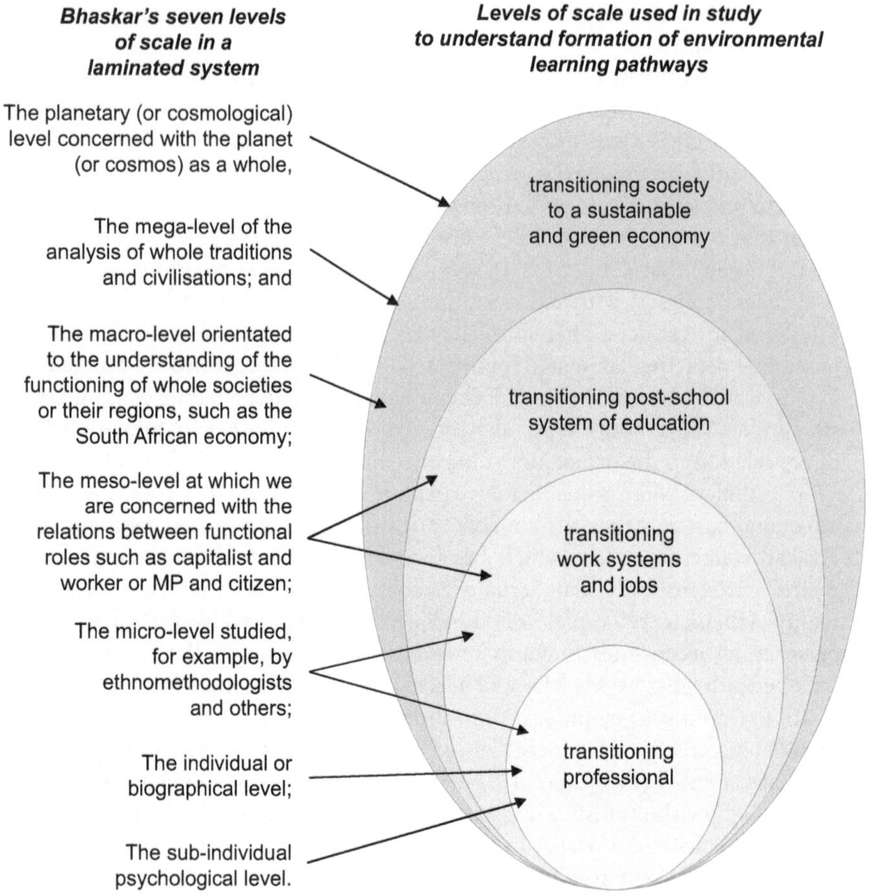

Figure 6.1 Reconstituting environmental learning pathways as a laminated system
Source: Ramsarup, 2017, p. 42

analysis of the operations and approaches of the professional associations). The scope of the data generated is also reflected in Figure 6.2.

To approach the analysis of the cases, Ramsarup developed a three-phase methodological process, outlined below.

Phase 1 – Career stories

In the first phase career stories were collected from 11 environmental scientists. The focus across the career story was educational progression into an environmental science job. These stories have tried to reflect both the processes and

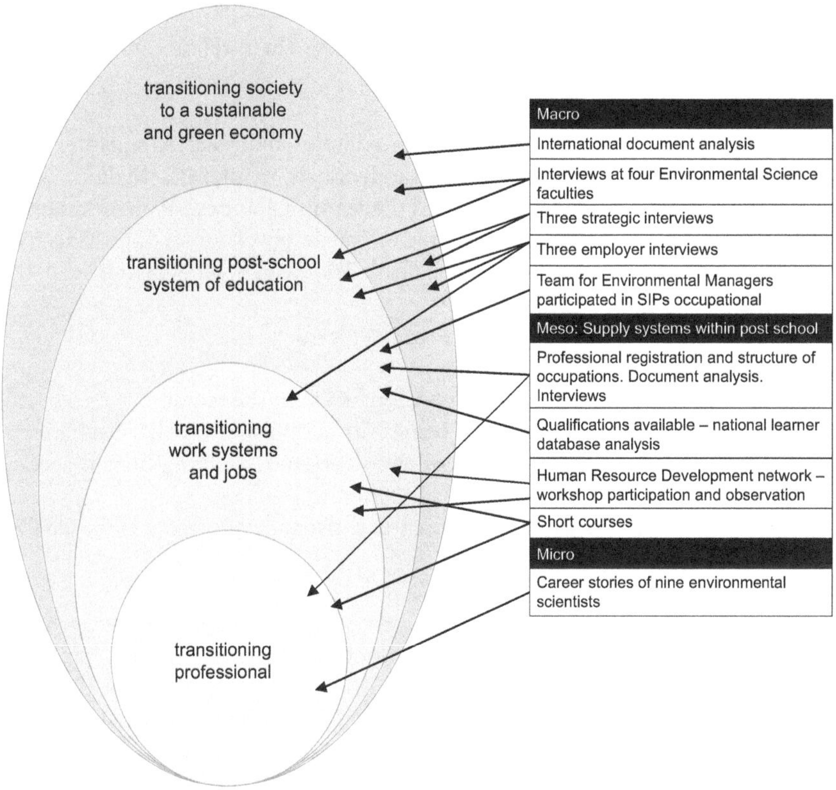

Figure 6.2 Scope of data across different levels

outcomes of transitions. Careers are inherently dynamic – evolving, developing and intersecting in planned and unplanned ways making it problematic to utilise linear, agentic frameworks for understanding career trajectories (Cohen & Mallon, 2001). Storytelling is a fundamental human activity and hence stories have emerged as a major source of understanding about how careers work. As people experience things and tell stories about their career experiences they also construct their personal realities (Patton & McMahon, 2006) and it is through these constructions of reality that we are able to see patterns and relationships between wholes and parts. Hence the stories have been recognised as useful, retrospective sense-making tools.

In substantiating the rich value of stories to career research, Cohen and Mallon (2001) highlighted four key related benefits:

- Sequencing – chronicle of events
- Revealing the inconsistencies and contradictions

- Retrospective sense making – gaze backward and invest past events with meaning that resonates with the present
- Enabling insight into how individuals view their relationship to social structures.

Stories enable the researcher to build a "rich, complex, multifaceted and integrated picture from the perspective of situated individuals" (Cohen & Mallon, 2001, p. 48). Ramsarup's study (2017) explored the learning journeys of environmental professionals "as the biographical narratives made up of inter-connecting sub-stories" (Black et al., 2010) told as their "path or trajectory through our available 'figured world'".

The career stories helped the study to:

- Capture the process notion of career formation in the sector
- Enable a focus on the 'human beings' in action and on the mechanisms underlying human action – this allowed the study to bring human agency and the dynamic nature of career formation to the fore
- Enable the capture of the relational and dynamic processes embedded in career development.

To identify consistencies and inconsistencies within and between career accounts the study utilised curriculum vitae (CVs) alongside the career stories. A CV is a rich source of longitudinal data, which lends itself well to the study of phenomena associated with careers and labour flows (Dietz, Chompalov, Bozeman, O'Neil, & Parker, 2000). The CV, as a data source, often recounts the entire career of the professional; it is thus not simply a list of credentials, but a historical document that evolves over time capturing changes in interests, jobs and collaborations, job moves and learning moves (ibid.). CVs are thus another chronological narrative form that represents the sequence and chronicle of career development. But despite the familiarity of this everyday artefact, knowledge of its social meaning, its research utility and the attendant practical problems in its research use is only beginning to accumulate (ibid). Inkson (2007) noted that stories can provide a commentary on external institutions affecting careers but little empirical work has been done within career research around this dimension.

Table 6.1 provides a brief introduction to some of the career stories collected. It illustrates the textured dynamics that emerged from the stories when they were analysed.

The career stories reflect many and numerous inter- and intra-organisational movements. The inter-organisational movements reflect movement into and out of mostly environmental consulting firms. Of the interviewees, only two have transitioned into parastatal organisations; the other career stories reflect mobility patterns between small and large consulting firms. Intra-organisational movement is linked to working on projects. The intra-organisational movement is largely prompted by roles within projects hence depicting largely lateral movements, with very little upward movement except when moving into a

Table 6.1 Environmental scientist career stories

Ref	*Story*
ES 1	Attempted a BSc but his school Physics was too weak and he struggled. He swopped into a BA Geography and Environmental Studies. Completed Honours in Geography and Environmental Management but failed to complete a master's degree. He started work in a small consultancy firm and developed a wetland interest and specialisation and skills in bio monitoring. He moved on to a large environmental company and progressed on to wetland divisional manager. He left as he started feeling his scientific technical skills were letting him down. He moved on to a smaller consultancy where he transitioned to working more in an environmental manager role. He subsequently moved on to a large environmental consulting agency as a project manager and is currently involved in managing large projects across Africa.
ES5	Geography was her favourite subject at school. Did a BA with a major in Geography. She progressed into honours in Geography because she thoroughly enjoyed Geography. During her honours she heard about the possibility of working as an environmental consultant. She started worked as an environmental consultant at Cleanstream and then joined an environmental consulting firm as a junior environmental scientist. She currently can't progress to senior environmental scientist as she is unable to register as an environmental scientist in SA.
ES8	Grew up with parents who were very keen gardeners and has grown up with a love for plants. He entered a BSc Botany and Zoology. On completion he needed to work so he didn't move straight into postgraduate studies. Was unable to secure a job as a black scientist so went to teach at a school. He applied for a lab assistant post, which he didn't get, but the professor gave him very good career advice and help. He then started as a nature conservator in a nature reserve at a university and worked part-time as a lab assistant in the Botany department. He left his family and travelled to work as a senior laboratory assistant in Zoology in a rural university. He spent six years away from home working as senior lab assistant and completed an honours degree. He returned home to work as a scientific officer in the herbarium in a botanical garden. Here he realised after three years that to work in research he needed to complete his qualifications and proceeded to complete an MSc and subsequently a PhD. He moved from being a scientific officer to a scientist and eventually to chief scientist within the herbarium.
ES6	Went to rural school in Limpopo. Understanding living things has always been her interest. She entered university but needed to do an extra year of academic support to improve her Science and Maths. She thereafter entered a BSc in Community Water and Sanitation Services. She was unemployed for one year and worked as a cashier at Postnet. She started honours in Environmental Management through correspondence. She then entered a graduate trainee programme in water and sanitation led by an engineering council. She worked as an industrial technician within the Department of Water Affairs. She then entered a three-year programme for young professionals. After this programme she was unemployed for a year. She thereafter was employed as a junior environmental project manager for five months from which she was retrenched. She was unemployed again for eight months and entered a job as environmental technician at a municipality.

more senior role within a project. The transition into a small consultancy meant interviewees were expected to work as a 'jack of all trades'. Thus, despite Geography and Environmental Management backgrounds, they were immediately expected to work within science roles; they also reflected that they were thrown into managing projects and specialising roles without much overall introduction and understanding of the broad scope of environmental science work. A further structural factor is the ability or inability to be recognised by the South African Council for Natural and Scientific Professions (SACNASCP) as a registered natural scientist, making professional status a critical structural factor influencing mobility. Three interviewees were unable to gain professional recognition, as the only registering body is SACNASCP and it does not accredit a Geography (social science) based environmental management route into the sector.

This study's contribution lies in connecting a deeper environmental career development perspective to the structures and systems that support education and training within environmental learning. In pondering the boundaryless career, which is relevant to the characterisation of environmental occupations, Weick (1996) maintained that careers consist of "work experiences that rise prospectively into fragments and fall retrospectively into patterns – a mixture of continuity and discontinuity" (p. 40). When we craft a story about our careers, we impose patterns with the benefit of hindsight. Inkson (2007) elaborated: "we know what we want in the present and that guides us retrospectively in our search for a plausible story" (p. 233). The critical realist underlabouring in the study helped to counteract some of the limitations of the retrospective data that emerged in the storytelling process.

Phase 2 – Institutional pathways mapping

Drawing again on the institutions supporting educational and occupational progression, in phase 2 the institutional structures supporting transitions of students within the environmental science and engineering occupations were investigated. Understanding the types of qualifications available remains a critical issue in understanding the official pathways available for potential students.

Information was also collected on the key aspects of supply linked to each occupation (qualifications available, defining of occupation, career systems, representation in national education and training systems). Figure 6.3 illustrates the elements of systems that were investigated.

Phase 3 – Systems mapping and absence analysis

Through constructing the learning pathways discourse as a complex multi-level phenomenon in this study, it was possible to surface various "internal contradictions and apporia" (Bhaskar, 2010, p. 21). Bhaskar explained the levels may be conceived as constituted by absence and omissions and various

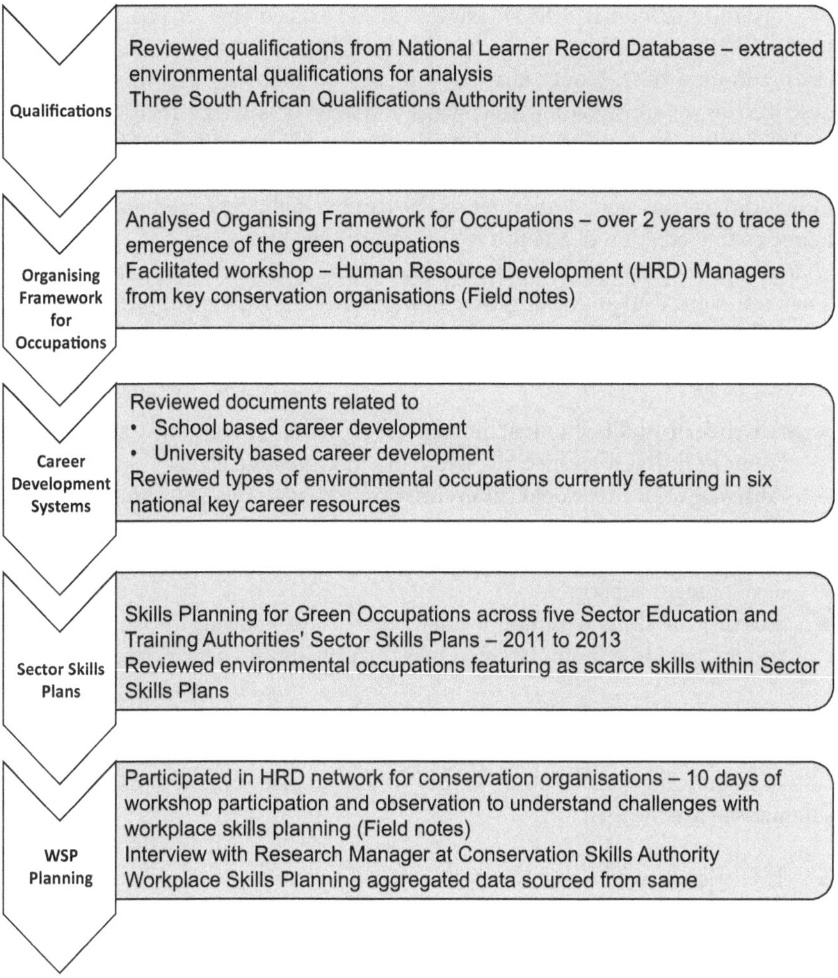

Figure 6.3 Mapping institutional pathways linked to environmental sciences

contradictions, as internally as well as externally related, as concrete and holistic and as involving processes dependent on transformative praxis understood as occurring on all four planes of social being, at up to seven orders of scale and as crucially dependent upon our discourse as a constitutive causally effected and efficacious feature of social life.

Across the career stories, as well as in the systems analysis, the environmental learning pathways studied were characterised by absences; these absences shaped the nature of the environmental learning pathways considered for the two occupations and were integral to the actualisation of the pathway. The career stories presented clearly illustrated how absenting could be viewed as a normalised practice within the crafting of environmental learning pathways.

Drawing on Bhaskar (1993), Nunez (2014) argued that envisaging holistic totality (being as whole) includes its splits, gaps (partial totalities) and its boundaries (sub-totalities). Nunez further outlined that mapping this totality can be described as a "constellation that maps a system of internal relations" (2014, p. 26). In this study, such a process encompassed the possibility of mapping a better learning pathway system for green work incorporating differentiated skills formation systems across all three skills bands (high, intermediate, low) and hence enable creation of a multiple entry, multiple exit system.

In the following section, four key absences identified in this study are discussed. They are central to unlocking the transformative praxis possibilities for skills development in the environmental sector.

These absences are:

- **Absence of qualifications** at different levels of the the National Qualifications Framework (NQF), especially entry and intermediate levels
- **Absence of occupational differentiation** in both skills level and specialisation and in terms of how occupations will change
- **Absence of comprehensive career development information** and career development support
- **Absence of skills planning mechanisms** or mechanisms to connect supply and demand in relation to green jobs within a futures perspective.

Absences can be considered in the MELD frame briefly outlined below. They are then examined in greater detail as interconnected within a laminated system frame. The four levels of MELD (Bhaskar, 1993; Shipway, 2010) can be summarised as follows:

- 1M – Being (what's real, actual)
- 2E – Absence (what's not there – causal power of absence)
- 3L – Totality (what could be there – reimagining totality)
- 4D – Transformative praxis (what are the possibilities? – how could it come into being?)

1M: *Data presented in as well as on career information illustrate that the dominance of high skill occupations and qualifications is perpetuated within career development systems. Available career information related to environmental careers is linked to widely known and traditional jobs. Furthermore, evidence presented in Figure 6.4 illustrates that career information is directed largely at the student transitioning from school into university.*

2E: *The data presented in the sections listed in Figure 6.4 clearly illustrates that information related to artisan, technical and low skill careers is absent from mainstream career systems like the National Career Advisory Portal (NCAP) as well as widely used environmental career guides. This absence is clearly interrelated with and emergent from the two absences presented above in relation to absences of qualifications and absences of broad occupational framing. Movement into specialist careers is predominantly*

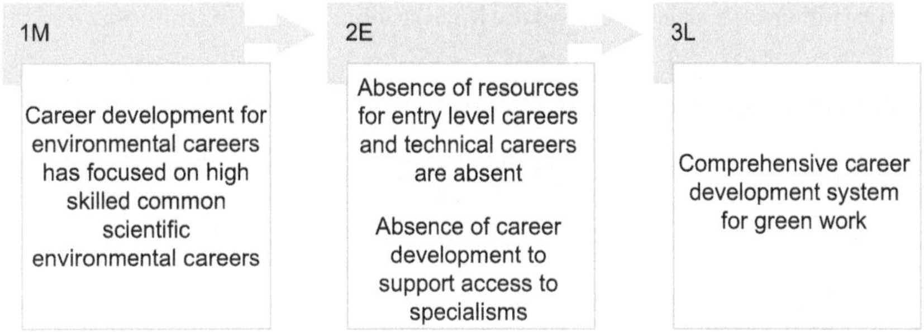

Figure 6.4 Absences in career development systems for environmental occupations

advocated through a master's degree (high skill qualification). This further supports the notion reported in several interviews that "to work in this sector you need a master's degree". Career information for graduates and people transitioning into specialised work is largely absent.

3L: *Clear signposts and high quality information are needed, as well as advice and guidance on green careers (that is well entrenched in national systems of career guidance like NCAP). With clear and coherent occupational information emerging across different levels of skill, career development information and guidance would be more inclusive, comprehensive and contain information for a broader selection of green work at entry, intermediate and high skill levels. Also, career systems will depict careers beyond science related fields to broaden social awareness of other types of environmental work and are able to present the full scope of environmental jobs.*

A prevailing observation is that access to environmental specialisms is reliant on the agentic processes of individuals who appear to be on a pathway that is not very transparent, lacks signposts and hence is characterised by non-linear progressions into environmental jobs that are not institutionally or occupationally determined. The dominant picture is thus of agents navigating their own non-linear pathway, moving in and out of education and paid work, participating in different projects and developing specialised knowledge and experience which enables their eventual transition into an environmental job.

Despite an obvious emerging demand, there appears to be a slow response from education and work systems. This illustrates that the national labour market intelligence system did not anticipate the demand for this emerging occupation.

The education and training system was unprepared and now the country is faced with trying to develop new skills with pathways for scientists that are reliant on the agentic action of individuals, without adequate systemic 'infrastructure' to respond to this substantive scarce skill need. This raises the issues of system immaturity and the inability to deal with 'newness'.

The analysis in this chapter demonstrates a reactive system lacking the ability for futures orientated demand insights, confirming findings of the Environmental

Sector Skills Plan study (DEA, 2010) that reported a largely reactive skills system to environmental concerns (cf. also the International Labour Organisation, 2011).

Conclusion and implications

Raffe (2003) has emphasised that to design effective policy we need to understand the real pathway so we can determine if the pathways on which policies are based correspond to the way pathways are actually experienced; if they are not, policies can be flawed. The evidence presented in this chapter has raised some interesting observations on how official maps of pathways are understood. Despite official pathway maps showing parallel ascending movements, the empirical career stories show more sideways navigations and flexibility. The lack of signposts may also increase the disadvantages of young people who lack the resources to make longer transitioning journeys.

The chapter has illustrated that understanding transitioning into environmental occupations necessitated the use of theoretical ideas from boundaryless careers discourse as well as critical realism which assisted in more accurately describing and depicting the nature of work and hence the transitional experiences of environmental professionals in these transitioning spaces. This has implications for the way in which supply issues are considered, the types of learning needs and courses, the needs of providers, the systemic articulation of educational and occupational pathways and how careers are enacted, as well as how education and training systems represent and conceptualise environmental learning pathways.

Acknowledgements

This study was part of a broader Rhodes University/South African Qualifications Authority (SAQA) research programme on Environmental Learning Pathways.

References

Arthur, M.B., & Rousseau, D.M. (Eds). (1996). *The boundaryless career: A new employment principle for a new organizational era*. New York: Oxford University Press.

Bhaskar, R. (1993). *Dialectic: The pulse of freedom*. London: Verso.

Bhaskar, R. (2010). Contexts of interdisciplinarity: Interdisciplinarity and climate change. In R. Bhaskar, C. Frank, G. Hoyer, P. Naess & J. Parker (Eds.), *Interdisciplinarity and climate change: Transforming knowledge and practice for our global future*, pp. vii–xiii. London: Routledge.

Bhaskar, R, & Danermark, B. (2006). Metatheory, interdisciplinarity and disability research: A critical realist perspective. *Scandinavian Journal of Disability Research*, 8(4), 278–297. doi: 10.1080/15017410600914329

Black, L., Williams, J., Hernandez-Martinez, P., Davies, P., Pampaka, M., & Wake, G. (2010). Developing a 'leading identity': The relationship between students mathematical identities and their career and higher education aspirations. Educational Studies in Mathemathics,73, 55–72.

Cohen, L., & Mallon, M. (2001). My brilliant career?: Using stories as a methodological tool in careers research. *International Studies of Management & Organization*, *31*(3), 48–68. doi: 10.1080/00208825.2001.11656820

Department of Environmental Affairs (DEA). (2010). *Environmental Sector Skills Plan for South Africa: A systems approach to skills planning and human capital development. Summary document*. (Draft 2). Pretoria: DEA.

Dietz, J.S., Chompalov, I., Bozeman, B., O'Neil, L., & Parker, J. (2000). *Using the curriculum vita to study the career paths of scientists and engineers: An exploratory assessment*. Atlanta: Georgia Institute of Technology.

Eby, L., Butts, M., & Lockwood, A. (2003). Predictors of success in the era of the boundaryless career. *Journal of Organizational Behaviour*, *24*, 689–708.

Fenwick, T.J. (2006). Escaping/becoming subjects: Learning to work the boundaries in boundaryless work. In S. Billett, T.J. Fenwick, & M. Somerville (Eds.), *Work, subjectivity and learning: Understanding learning through working life* 1st ed., pp. 21–36. Dordrecht: Springer.

Forrier, A., Sels, L., & Steynen, D. (2009). Career mobility at the intersection between agent and structure: A conceptual model. *Journal of Occupational and Organisational Psychology*, *82*, 739–759.

Gunz, H., Evans, M., & Jalland, M. (2000). Career frontiers: New conceptions of working lives. In M. Peiperl, M. Arthur, R. Goffee, & T. Morris, *Careers in a "boundaryless" world*, pp. 24–53. Oxford: Oxford University Press.

Human Sciences Research Council (HSRC). (2009). *Guidelines towards a human capital development strategy in the biodiversity sector*. Research report, The Lewis Foundation and the South African National Biodiversity Institute (SANBI). Pretoria: HSRC.

Inkson, K. (2007). *Understanding careers: The metaphors of working lives*. Thousand Oaks, CA: Sage. doi: 10.4135/9781483328867

International Labour Organization (ILO). (2011). *Skills and occupational needs in green building*. Retrieved from www.ilo.org/wcmsp5/groups/public/---ed_emp/---ifp_skills/documents/publication/wcms_166822.pdf

Lindley, D. (2014). *Expansive social learning in wetland management: A case study of the Mondi Wetlands Programme*. Unpublished PhD thesis, Rhodes University, Grahamstown, South Africa.

McKenzie, P. (2000). Youth pathways to promote lifelong learning. Presented at the Centre for the Economics of Education and Training (CEET) Conference 'Mobilising resources for lifelong learning', Ascot House, Melbourne. Retrieved from http://citeseerx.ist.psu.edu/viewdoc/download?doi=10.1.1.549.5575&rep=rep1&type=pdf

Nunez, I. (2014). *Critical realist activity theory: An engagement with critical realism and cultural-historical activity theory (CHAT)*. London: Routledge.

Patton, W., & McMahon, M. (2006). *Career development and systems theory: Connecting theory and practice*, 2nd ed. Rotterdam: Sense Publishers.

Price, L. (2012, March). *Interdisciplinarity and critical realist research*. Paper presentation at the PhD Week, Education Department, Rhodes University, Grahamstown, South Africa.

Raffe, D. (2003). Pathways linking education and work: A review of concepts, research, and policy debates. *Journal of Youth Studies*, 6(1), 3–19. doi: 10.1080/1367626032000068136

Ramsarup, P. (2017). *A critical realist dialectical understanding of learning pathways associated with two scarce skill environmental occupations within a transitioning systems frame*. Unpublished PhD thesis, Rhodes University, Grahamstown, South Africa.

Roper, J., Ganesh, S., & Inkson, K. (2010). Neoliberalism and knowledge interests in boundaryless careers discourse. *Work, Employment and Society*, 24(4), 661–679. doi:10.1177/0950017010380630

Sayer, A. (2000). *Realism and social science*. London: Sage.doi: 10.4135/9781446218730

Shipway, B. (2010). *A critical realist perspective of education*, 1st ed. London: Routledge. doi: 10.4324/9780203881880

Smith, R. (2011). Evolving higher education. Environmental scientist. *Journal of the Institution of Environmental Sciences*, 20(2), 26–33.

Tams, S., & Marshall, J. (2011). Responsible careers: Systemic reflexivity in shifting landscapes. *Human Relations*, 64(1), 109–131. doi: 10.1177/0018726710384292

Vaughan, K. (2003, August 8). *Changing lanes: Young people making sense of pathways*. Presented at the New Zealand Council for Educational Research (NZCER) Annual Conference, 'Educating for the 21st Century', New Zealand. Retrieved from www.nzcer.org.nz/system/files/12223.pdf

Weick, K.E. (1996). Enactment and the boundaryless career: Organizing as we work. In M.B. Arthur & D.M. Rousseau, *The boundaryless career: A new employment principle for a new organizational era*, pp. 40–57. New York: Oxford University Press.

7 Transitioning into work

A learning and work transitioning process perspective

Presha Ramsarup and Heila Lotz-Sisitka

Introduction

Across the pages of this book we have indicated that social-ecological issues create new challenges for work and learning research. This is similarly the case in understanding transitions into green work, from both intra- and inter-institutional context perspectives. In this chapter, we consider socio-ecological challenges that are shaping the demand and articulation of scarce skills, such as the expansion of the wildlife economy leading to the wildlife veterinarian being identified as a scarce skill at national level. We also consider sustainable development challenges (e.g. provisioning of clean water, waste management systems and functioning sanitation management amongst others) in local government contexts leading to a need to better understand the roles and transitioning opportunities for employees in local government contexts. We first review the landscape of learning and work transitioning research and identify an important gap in this research, which is to better understand learning and work transitioning experiences and processes, framed within a discourse of critical vocationalism, where emancipatory intent shapes the discourse. This brings greater attention to structural as well as agentive factors in transitioning processes, and the ways in which these are interrelated. This in turn creates methodological challenges for learning and work transition researchers. We show two case examples of methodological approaches developed by researchers addressing the challenges noted above via their learning and work transitioning research.

Against the background of an emerging sector (cf. Chapter 1), this chapter uses a lens of learning and work transitions, applied to environmental learning pathways to explore the complexities associated with pathway construction, articulation and efficacy in systems of provisioning.

Learning and work transitioning discourse and research

A number of authors have recognised that various professions are experiencing a heightened sense of transitions (Evetts, 2009; Fenwick, 2013; Sawchuk & Taylor, 2010). Field (2012) reflected that the boundaries and expectations of learning and work transitions though the life course are changing, both at the level of the individual and at the level of the wider society. An Organisation for Economic Co-operation and Development report (OECD, 2008) noted that learning and work transitions are increasingly fragile and exclusionary and that the number of 'stepping stones' required by youth to secure a labour market position is multiplying. Within this, individual learning and work transitions are becoming more challenging (Sawchuk & Taylor, 2010) and little is known about these transitions beyond transitioning into the first job. While there is a considerable amount of literature focusing on transitions into first jobs, a gap exists in the literature on learning and work transitioning in the context of specific scarce skills, which we aim to address in this chapter via the case study below on wildlife veterinary scientists (cf. Mckrill, 2015; see also Chapter 6 for insight into this). Additionally, there is little literature on transitioning within workplaces, which we address here via the case study on transitioning within a local government workplace context at different levels of sustainable development occupations (Mohanoe, 2013).

We address these gaps because individuals are being called on to manage transitions throughout their careers – beginning with the shift from initial professional education to workplaces. Because learning and work transitions are becoming more challenging due to the changes in economy and ecological challenges, economic structure, precarity around work and an amplified tendency towards specialisation associated with modern forms of work, especially in the area of high skills, there is a critical need for education and work systems to pay more attention to *experiences of transitions* (i.e. transitioning processes) in a diversity of transitioning contexts (cf. Chapters 6 and 8).

As already argued in Chapter 6, the supply and demand approach to understanding learning pathways in a South African context within the environmental sector was insufficient; hence, trying to understand the transitions people were making became an important lens in constructing the real nature of the learning pathways. In this chapter, we develop this perspective further by offering more insight into methodological ways of researching experiences of learning and work transitions. In addressing the limitations of supply and demand discourse, we found the notion of *transitions* to be an important concept for understanding wider systemic articulation concerns, as it provides both structural and systemic insights (cf. Chapter 6). Our research shows that studies of transitioning experiences provide empirical insights into how various qualifications may or may not articulate, and how the changing context shapes choices and needs related to the available qualifications, how they are used and how they relate to more traditional qualifications-led articulation of learning pathways (cf. Chapter 6; Ramsarup, 2017).

Eccelstone (2009), writing with a focus on *individual transitions*, agreed that transition could involve movement but argued that it is much more than this. She proposed that transition was a "change process but also a shift from one identity to another" (p. 125). Life course researchers, like Eccelstone, argued that effective transitions require us to understand how people progress cognitively, emotionally and socially at different stages and how they navigate the complex demands of different contexts (p. 126). Eccelstone (2009) summarised her ideas into three key meanings of transitions: 1) navigating pathways, structures and systems, 2) becoming someone where transition is between two states of being and 3) life as transition where the whole of life is a form of transition. All this points to the idea that mainstream notions of supply and demand used in the skills system to analyse entry and exit into jobs are insufficient and lack viability to understand the transition into jobs undergoing change, hence also our emphasis on *transitioning experiences* which surface insight into *transitioning processes*, offering more nuanced perspectives for transitioning research. As argued below via the case studies, this also requires methodologies that allow for surfacing transitioning experiences, and thus generate perspective on transitioning processes.

Raffe (2008) provided a useful frame for helping to develop an understanding of the *macro aspects of transitions*, which he outlined as different institutional and structural arrangements that shape young people's education-work transitions. This work enabled us to identify components that require attention and it provided a useful way to conceptualise the notion of a 'transition system' as a multi-dimensional concept that involves strong interdependent relationships between education and the labour market and societal, historic and political factors (cf. Chapters 1, 3, 4, 5, 6). It helped us to understand the components within institutional arrangements and how they can impact and shape transition processes and outcomes.

Notions of systemic/institutional pathways and individual pathways have emerged across the literature as indicated in the discussion above, and the dualism described widely characterises the literature within learning and work transitioning. Sawchuk and Taylor (2010), however, mapped the landscape of transitional research by drawing out a continuum of research shaped by 'new vocationalism' leaning their research differently towards the other end of the continuum, namely 'critical vocationalism'. Here they stressed that new vocationalism (which draws strongly from economic approaches) and critical vocationalism (drawing strongly from critical theory) remain on a continuum (so as not to create two 'camps'). They argued (ibid.) that the central underpinning idea in new vocationalism is that of 'human capital', where workers need to adapt to the needs of the economy, thus foregrounding an individual/adaptive perspective. This, they argued further, has inherent difficulties in admitting structural contradictions rooted in both institutions of education and the economy, which we argue is an important dimension of green skills research (cf. Chapters 1 and 6).

Transitioning research imbued with the underlying propositions of new vocationalism minimises attention on how education systems are integrally related to labour markets and work systems and fails to fully analyse how these transitions are "shaped by the tensions and contradictions inherent within processes of control, conflict, accommodation and occasionally resistance" (Sawchuck & Taylor, 2010, p. 9). Responses from within a vocationalist mindset tend to assume that more responsive education, continuing education and vocational and workplace training (from cradle to grave) would address the problems in the economic labour market and wider social-ecological system contradictions (e.g. those influencing emerging green skills). However, as illustrated by Livingstone (2004) and Sawchuk and Taylor (2010), this remains a flawed thesis, as a significant 'education-jobs' gap remains under-recognised and misinterpreted within the new vocationalist position (Sawchuk & Taylor, 2010, p. 9).

Critical vocationalism has its theoretical roots in areas of critical theory applied to work contexts that are concerned with race, disability, social class, gender (and we can add ecological degradation, cf. Moore, 2015). This body of work represents a school of thought intended to disrupt the status quo. Sawchuck and Taylor (2010) argued that central to the critical vocationalist perspective is an understanding of context, social differences and power relations that "define[s] how learning capacities are productive and reproductive of uneven social and economic prosperity" (p. 1). Their argument is that these need to be emphasised in transitions research. Bynner (2001, as cited by Sawchuk & Taylor, 2010), in reflecting on learning and work transitions, highlighted the "interconnectedness of activity across the different domains of life" (p. 9), further emphasising that there is a lack of attention on the broader political climate, culture and effects on other spheres of institutional and non-institutional life as dimensions of the transition process. This has a significant bearing on our interest in articulating learning and work transitioning pathways in a green skills context, where social-ecological and political-ecological conditions significantly shape this emerging area of skills research (cf. Chapters 1, 2, 3, 4 and 5).

The perspectives derived from critical vocationalism and the associated implications for transitioning research highlighted above resonate very strongly in our studies on learning pathways (Lotz-Sisitka, Ramsarup, Olvitt, & Mohanoe, 2017; Mckrill, 2015; Mohanoe, 2013; Ramsarup, 2017, cf. Chapter 6) and confirm that educational pathways and qualifications are not neutral mechanisms for allocating opportunities to people. There is much more that needs to be understood from a critical analysis of learning pathways' experiences, especially in a context like South Africa where deep inequalities pervade workplaces, and access to qualifications affect learning pathways and therefore also transitioning experiences and possibilities.

In addition to calling for more attention to social variables, Sawchuk and Taylor (2010) argued for an expanded view of learning and work transitions that encompass *inter-institutional* boundary crossing between formal and non-formal learning experiences as well as *intra-institutional* learning and work transitions that are focused on transitioning into different types of work, sectors and occupations. The two case studies below address these: inter-institutional boundary crossing (cf. Chapter 6) and intra-institutional transitioning.

Sawchuk and Taylor (2010) argued that an important theme in critical vocationalism is informal dimensions of transitions, learning and experience, which are useful in studies on environmental learning pathways. They claimed that to understand the range of complexity of learning and work transitioning throughout the life course, we need to recognise "learning and experience throughout its full range of variation" (p. 1), which they posited is absent from dominant theoretical and policy based research. The neglect of informal learning, they contended, is due to the difficulty to credit and commodify it. The critical vocationalist approach helps to highlight these barriers and their impact on individuals and groups.

However, while these important points are raised for learning and work transitions research, not enough is said about the methodologies that can be used to surface such analysis. It is this that we address in the two case studies below. As in the other chapters, our research is informed by and underlaboured with critical realism. In both of the studies below, the research drew mainly on Basic Critical Realism's capability to uncover depth structural dynamics and generative mechanisms shaping experiences of learning and work transitioning (cf. Chapter 13), thus also illuminating important dimensions of these transitioning processes that resonate with the critical vocationalist end of the continuum proposed by Sawchuk and Taylor (2010).

Case studies examining transitioning

Wildlife veterinary scientists: Researching inter-institutional learning and work transitioning experiences

The study of Mckrill (2015) offers a good case of how to research inter-institutional learning and work transitioning experiences in the environmental sector. The study centres on the higher education to workplace transitioning of wildlife veterinary scientists (wildlife vets) in South Africa. Wildlife vets were identified as a scarce skill in biodiversity sector skills demand research (SANBI & The Lewis Foundation, 2010).

Supply research indicated that the only available training programmes for wildlife vets was at the University of Pretoria Veterinary Scientist Specialist Training Centre, Onderstepoort. Demand studies were showing a high need for wildlife vets. Yet very little was known about the supply dynamics, or the demand dynamics creating the scarcity of wildlife vets in South Africa. Mckrill (ibid.), at the start of her study, found that almost nothing was known about the learning and work transitioning experiences of wildlife vets in South Africa, despite this being recognised as a scarce skill in a burgeoning industry.

To address this problem, Mckrill (2015) needed to design a study which would provide insight into the supply side dynamics, as well as the demand side dynamics, and which would also give her insight into the experiences of wildlife vets in transition from higher education into the employment sector.

She used qualitative research approaches to interview a number of people in the biodiversity sector who were involved in the supply and training of wildlife vets, people who were articulating and experiencing demand for wildlife vets and wildlife vets who were in training and/or who had or were in transition from higher education into the workplace. She also gathered documentary records and reports which offered perspective on these dynamics. Interviews with wildlife vets were determined by the supply side dynamic, i.e. only one elective specialism course for wildlife vets was on offer at Onderstepoort, namely the Wildlife, Ostrich and Crocodile 6th Year 10th Semester (WOC 610) course. In interviewing trainees in this programme, she included questions on their perceptions of the course, as well as their ideas pertaining to future learning and work transitions and the factors that could affect them (Mckrill, 2015, p. 87). She also tracked why there were such limited training opportunities for wildlife vets (via historical interrogation) and identified that it was the early twentieth century expansion of domestic livestock populations as the agricultural economy grew that led to an absence or neglect of Wildlife Veterinary courses. Thus, most wildlife vets specialised in the USA or UK due to a lack of training opportunities for this area of specialism in South Africa until the 1980s when the WOC 610 course was introduced at Onderstepoort.

On the demand side, in earlier times, the wildlife sector had low economic value compared to domestic livestock. Game animals were heavily hunted, until their populations were seen to be under threat, which led to the establishment of national parks and conservation institutions. Since 1994, the wildlife economy has grown exponentially, especially in the areas of game ranching and game farming trades (ibid.), expanding to 8 billion ZAR per annum in value (Wildlife Ranching South Africa, 2012). By 2010, wildlife vets were in high demand and were identified as a scarce skill. Mckrill (2015) uncovered the main reason for this is the legislative driver from the Medicines and Related Substances Control Act of 1965 (Act 101 of 1965) (Department of Health, 1965) that requires a vet to be present when wild animals are darted or moved as the drugs used are Schedule 5 or 6 drugs. While there is a high demand for wildlife vets due to this change in the economy and legislation surrounding wildlife management, Mckrill's study reports that in 2015 the number of veterinarians entering the wildlife industry remains unknown and that there is no funding from government to investigate this area of skills demand (p. 90). This indicates a mismatch between skills planning, demand, analysis and provisioning in the biodiversity sector.

The supply and demand dynamics revealed above via contextual and historical analysis were crucial for bounding or framing the more detailed study of learning and work transitioning experiences, and helped Mckrill (2015) to focus on students in the WOC 106 course, academics and employees familiar with the older conservation and the new game ranching contexts. Mckrill was particularly interested in identifying enabling or constraining factors that were influencing her research participants' transitions. To generate as broad as possible a view on these, she drew on Urie Bronfenbrenner's (1979) human ecological systems theory to guide analysis of the data, which she underlaboured with critical realism. This choice of theoretical framework also influenced the scope and range of questions she asked in interviews, but more importantly it offered her a means of analysing learning and work transitioning experiences and associated systemic and other social processes influencing these, as recommended by Taylor and Sawchuck (2010).

Bronfenbrenner's (1979) human ecological systems theory focuses on a nested set of interconnected 'systems' not unlike the laminated totality framework of Bhaskar (cf. Chapter 6, Tikly, 2015). It also explains that an individual's experiences are influenced by environmental layers within an interrelated

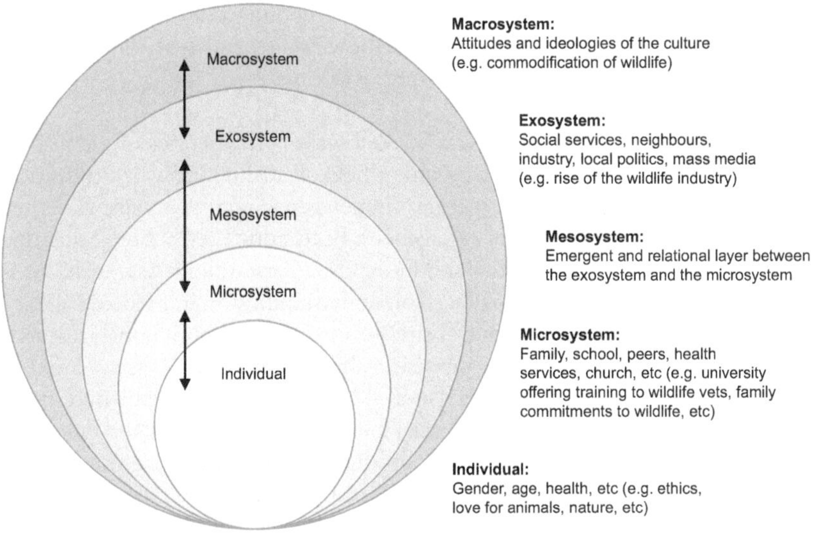

Figure 7.1 Bronfenbrenner's human ecosystemic model
Source: Adapted from Bronfenbrenner, 1979

system, with Bhaskar (2010) giving greater emphasis to a reality outside of our knowledge than Bronfenbrenner. Bronfenbrenner's (1979) model centres around the individual, and he uses the terms, microsystem (family, school, church, peers), mesosystem (interconnections between microsystems and the individual), exosystem (neighbours, social services, industry, local politics, media), macrosystem (attitudes and ideologies of culture) and chronosystem (patterning of environmental events and transitions over time/the life course) to describe his laminated ecosystemic model as shown in Figure 7.1.

According to Mckrill (2015, p. 47), Bronfenbremmer explains "an ecological transition as a process occurring across the aforementioned systems, noting that when an individual's position is altered in the ecological environment, it will involve a change in setting or role or both". Systemic aspects, structures and roles in this model possess emergent properties that have power to alter ways of thinking, acting and feeling. However, people's agency in defining or co-defining their roles as they transition from one role (e.g. student) to another role (e.g. wildlife vet) cannot be discounted, and here critical realism offers the transition researcher tools that help to understand the dynamic interplay between structural influences from the past on the agency of a person to act and make decisions along a pathway (or not), and how the individual persons' agency or ability to act (alone or with others) influences the formation of the learning and work transition pathway. Eccelstone (2009) noted that it is difficult to separate agentive factors from structural influences. Responding to this problem in the social sciences, Archer (2000), drawing on Bhaskar's Transformational Model of Social Action (TMSA), proposed using the methodological strategy of analytical dualism to 'separate out' (analytically) structural, cultural and personal emergent properties as this helps one to see how agency emerges from structure, and how agency can also change structure over time (see Lindley (2014) for a good example).

Both Bronfenbrenner and Bhaskar worked with both theories of a stratified ontology and theories of emergence which helps learning and work transition researchers to understand how structures shape agency and vice versa (i.e. they can get insight into the process of emergence). Ecclestone (2009) noted that one of the reasons for limited literature and in-depth understandings of transitions is that too much emphasis has been placed on understanding an individual's agency and identity in learning and work transitions research, and not enough research considers structural factors and how they shape agency and vice versa (cf. Sawchuck & Taylor, 2010). For this reason, Mckrill (2015) worked with critical realism, as it helped her to "identify casual mechanisms, structures and emergent properties, to counter the over-emphasis on the individual narrative and identity in transition research, while not neglecting these" (p. 43). Ecclestone (2009) went on to say that transition researchers should focus their energy on "easing the process of transitions via structural interventions and mechanisms" (Mckrill, 2015, p. 43). However, Fenwick (2013) warned not to interpret transitions in a linear manner, noting that there is unpredictability, diversity and multiple pathways to take account of in learning and work transitions research, and there are also influences from wider 'transition systems' to account for (cf. Chapter 1).

In her analysis, Mckrill (2015) first analysed the data for enabling and constraining factors using an inductive approach to analysis, identifying factors such as passion or motivation, networking, family influences, mentors, finances (enabling factors), and limited opportunities, lack of subject knowledge, responsibilities, etc. (constraining factors). Following this, she engaged in what she called secondary analysis, using an abductive mode of inference (cf. Chapter 13), using Bronfenbrenner's model to recontextualise or interpret the data as follows:

> Firstly I focused on the microsystemic factors influencing an individual's transition, namely exposure and experience within the biodiversity field, as a result of any microsystemic relationships they held with family members or friends. Secondly, I analysed macrosystemic factors, within both interviews and document analysis. Thirdly and lastly I analysed and sought to identify the exosystemic and chronosystemic factors which indicated external, often structural factors which influenced an individual's development and greater transition.
>
> (p. 78)

Some of the findings from Mckrill's (2015) study show what this analytical framework affords the green skills researcher interested in learning and work transitioning experiences. For example, it revealed that early microsystemic factors such as parental or friendship influences during childhood, and exposure to the field (exposure to wildlife and animal husbandry) contributed to an interest in becoming a wildlife vet. Macrosystemic factors such as the availability of good quality schooling and relevant university degree courses also shaped transitioning into becoming a wildlife vet, especially the availability of the WOC 160 elective course. Here, macrosystemic and broad chronosystemic elements such as apartheid which left poor quality education in its wake, and which left many without adequate resources/scholarships for study, also shaped the learning and work transitions of the wildlife vets, while issues such as a lack of available supervisors influenced the abilities of students to specialise as wildlife vets.

True to its interest in revealing ontological depth, critical realism underlabouring helped Mckrill (2015) to identify generative mechanisms using a retroductive mode of inference (Price, 2017, cf. Chapter 13). Through this, she was able to identify generative mechanisms shaping these learning and work transitional experiences. For example, she discovered it is the expanding commodification of wildlife which is partially causal in the changing nature of veterinary work, influencing the above transitioning experiences. Another example of a generative mechanism identified is the risk of disease transmission from wildlife to domestic livestock, which shaped the dominant emphasis on domestic livestock veterinary practice at the training institute, leading to limited options for qualifying as a wildlife vet. As such, critical realism helped Mckrill to "better describe the relations between the whole and the parts" (2015, p. 50) (cf. Chapter 13). It also helped her, as a work and learning transitions researcher, to address the recommendations of Sawchuck and Taylor (2010) and Ecclestone et al. (2009) that skills system researchers should focus their energy on "easing the process of transitions via structural interventions and mechanisms" (cf. above). In this case, it is clear that issues associated with the commodification of wildlife as well as quality and availability of educational opportunity are two spheres where interventions are needed to address the issue of the wildlife vet as 'scarce skill'.

Municipal sustainable development employees: Researching intra-institutional learning and work transitioning experiences

A second case, this time focusing on researching intra-institutional learning and work transitioning experiences, conducted by Mohanoe in 2013 into the learning pathways of workers, managers and supervisors in a local municipality, offers further methodological possibilities for researching learning and work transitioning experiences. In this case, Mohanoe (2013) sought to understand how training opportunities allow for intra-institutional learning and work transitioning experiences. Designing such a study required her to find out how individuals in key occupations in the Municipality have followed, follow or would like to follow learning pathways related to sustainable development by focusing on the past, the present and the future of employees' training experiences, and related issues of articulation, progression and access within a systems perspective in this context. She therefore framed her research as an in-depth case bounded by this scope of enquiry.

She was particularly interested in the issue of articulation, and how this shapes intra-institutional learning and work transitioning, citing that

> the principle that makes an integrated system come to life for the benefit of all learners is that of articulation (i.e. the actual connections between different learning pathways at the horizontal level and transitions from one level to the next in vertical terms).
>
> (ADEA, 2008, p. 69)

As noted by Lotz-Sisitka and Ramsarup (2011), "the question of articulation is tied to quality; and to transfer related questions across the National Qualifications Framework (NQF) sub-frameworks; and to emergence in the context of learning pathways. It assumes boundary zones; and transfer processes (ibid.)".

This drew attention to the need to examine articulation from vertical and horizontal perspectives in learning and work transitioning research. This led Mohanoe (2013) to focus on boundary crossing across the different sub-sectors of the NQF, namely from the Umalusi governed schooling sub-sector, to the Qualifications Council for Trades and Occupations (QCTO) governed accredited workplace learning sub-sector and the Council of Higher Education (CHE) governed higher education qualifications sub-sector. In intra-institutional transitioning, all of these sub-sectors of the NQF are relevant, especially in contexts such as municipalities where a large percentage of the workers are in the elementary occupation level, with generally poor levels of prior education (i.e. many have not finished basic education qualifications). Thus there is a need for adult education provisioning. There is also need for vocational education and training to address specialist technical needs, such as water quality monitoring. Such training can be formal (i.e. obtained before employees join the municipality from technical and vocational education and training (TVET) colleges or universities of technology) or workplace-based (i.e. offered in the workplace via accredited skills programmes). There is also a need for high skills such as specialisation in Environmental Management, with such upskilling often taking the form of part time or full time studies in higher education institutions. All of this influences possibilities for intra-institutional transitioning in sustainable development occupations.

To study this, Mohanoe (2013) focused on investigating articulation and access issues in selected learning pathways to consider their location within and across the sub-frameworks of the NQF. In this she focused further on the issues influencing crossing from one area to the other, blockages, tensions and enabling factors allowing learning pathways for key occupations. This drew on research by Lotz-Sisitka and Ramsurup (2011) which identified 'researching the boundary' as an important unit of analysis for learning pathways research:

> a unit of analysis focusing on the 'cross overs' in learning pathways research can provide a richer picture of 'learning pathways'. Such a unit of analysis potentially provides a more interesting way of understanding articulation in the context of learning pathways; albeit it being more complex.

To gain depth insight into transitioning experiences, she sought to understand individual experiences of employees in key occupations at elementary, supervisory and management levels. She probed their experiences of learning pathways using career stories (cf. Chapter 6) which enabled her to explain why individuals make the choices they make, and the meaning that guides these choices and decisions. She noted that such stories have the following benefits: they provide a chronicle of events, they reveal inconsistencies and contradictions, they provide perspective on the past, present and future and they enable insight into how individuals view their relationship to social structures (Mohanoe, 2013, p. 58, citing Cohen & Mallon, 2001), allowing the researcher to build a "rich, complex, multifaceted and integrated picture from the perspective of situated individuals" (ibid.). Integrating policy influences and contextual and practice analysis with interview and focus group data, Mohanoe (2013) composed career stories for each of her selected research respondents, constituting a first layer of analysis. These career stories revealed various intra-institutional learning and work transition experiences and dynamics including, but not limited to:

- A lack of basic education created blockages for workers in terms of mobility and progression in sustainable development occupations
- A lack of adequate socio-economic support significantly affected learning and work transitioning starting points, as well as capabilities to make use of transitioning opportunities
- Lack of progression options in learning and work transitions from casual to permanent workers
- Personal motivational factors such as lack of interest, or high levels of motivation for progression influenced learning and work transition experiences and processes
- Poor workplace skills planning, coupled with absence of training opportunities such as Adult Basic Education and Training (ABET) programmes or bursaries for further learning significantly influenced transitioning of employees at all levels, but especially those at elementary occupation level whose power was less evident for agency-driven professional learning than those employed at higher levels in the institution
- Poor worker-manager relations and lack of support from managers and/or supervisors at different levels in the system affected learning and work transitioning experiences significantly
- Material factors such as inadequate financial resources, perceived corruption practices and lack of proper working equipment also hampered intra-institutional learning and work transitioning.

See Mohanoe (2013) and Lotz-Sisitka et al. (2017) for further insights into these and other aspects influencing learning and work transitioning 'at the boundary'.

All of these affected effective service delivery, highlighting the important relational dimension of learning and work transitioning in an intra-institutional context. Also informed by the Taylor and Sawchuck (2010) and Eccelstone et al. (2009) arguments for a stronger engagement with critical vocationalism and a need to focus attention on addressing structural processes that affect individual (and in this case types of employees – workers, supervisors, managers) learning and work transitioning pathways – Mohanoe (2013) deepened the career story

Table 7.1 Generative mechanisms shaping intra-institutional transitioning of workers, supervisors and managers in sustainable development learning pathways in a local municipality

Workers	*Supervisors*	*Managers*
1. Inequality 2. Social relations and lack of freedom (i.e. oppressive power relations) 3. Marginalisation and lack of training provided 4. Organisational inefficiency 5. Political neglect 6. Legislative requirements and compliance 7. Gender issues 8. Language 9. Communication	1. Access to education 2. Choice of subjects 3. Higher education and gender issues 4. Marginalisation 5. Retirement age 6. Legislative requirements	1. Access to higher education 2. Access to resources 3. Legislative requirement 4. Organisational inefficiency

analysis with a critical realist generative mechanism analysis. Like Mckrill (2015), she did this via retroductive inference (cf. Chapter 13) from the career stories, depicted in Figure 7.2. Interestingly, she was able to identify different generative structural mechanisms shaping the transitioning experiences and learning pathways of the three types of employees (see Table 7.1), showing a 'heavier impact' of these mechanisms on those with least institutional power to direct their learning pathways, namely the elementary occupational workers.

This study pointed to the urgent need to address power imbalances and political neglect and marginalisation of elementary occupation level workers in the intra-institutional learning and work transitioning context. This in turn required giving attention to power relations in the workplace, organisational efficiency, adequate workplace skills planning and other interventions needed to address this critically important generative mechanism, illustrating the importance of considering the recommendations of Sawchuck and Taylor (2010) as well as Eccelstone et al. (2009) for learning and work transitioning research, especially transitioning research orientated towards emancipation and the realisation of sustainable development objectives. This finding is important more widely for addressing problems of service delivery in municipalities in South Africa, and for addressing the ongoing legacies of exclusion, including how exclusions continue to manifest, even in local government contexts, despite the advent of democracy in 1994 and intentions to address these historical legacies. The study also shows the vital importance of functioning skills systems in local government contexts for advancing sustainable development from a social equity point of view, as well as from an ecological and economy praxis point of view.

Causation diagram – workers

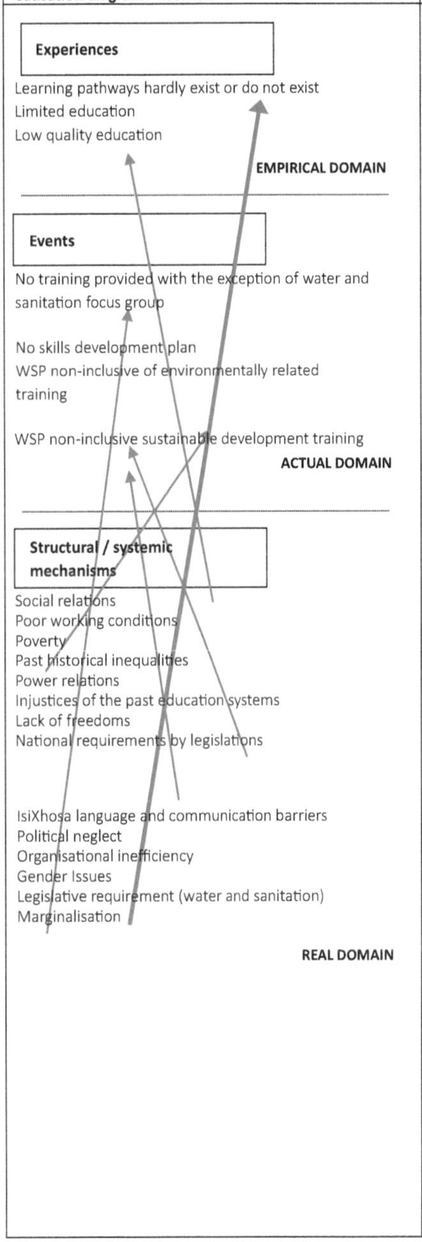

Retroductive explanation

The workers have either non-existing or hardly existing learning pathways in their key occupations. This is shown in their experiences of low quality and limited education. No training is provided for them either relevant to the environment or their occupations. Out of the nine focus groups investigated, the only training provided was for the water and sanitation group. This is the case because of the National legislation from the Department of Water (DWA) stipulates that training is imperative to all water related occupations, and the DWA was funding such training. This is also a skill identified as scarce and critical and therefore prioritised for training at all levels in the environmental sector and local government.

There is no or very poor skills development planning in place for sustainable development implying actualization of planning of skills, no incorporation of relevant and integrated skills training for the workers. Despite guidance available in the Environmental Education and training strategy, the Workplace Skills Plan (WSP) in place does not make provision for accommodating environmentally related training and neglects the workers critical training. The unravelling of underlying systems and structures which cause the non—existence and hardly existing learning pathways of the workers indicate various mechanisms at play, in interaction. These show reality, which can never be pre-determined or preconceived, and that many structural mechanisms need to be engaged at the same time for change to occur.

They include inequality of past historical injustices on the education system which has perpetuated the situation, social economic factors such as poverty and unemployment in extremes being barriers to learning pathways, tensions of power relations between workers and managers emanating from the unequal economic structuring of work in modernity. This causes blockages, marginalisation and the Municipality's organisational inefficiency in general affecting processes, decisions on training ultimately neglecting the workers' critical needs, compulsory legislative requirements enforcing training where provided, gender stereotypes on choice of training unrelated to occupation, constraining the emergence of successful training. Language and communication were also identified as causal factors impeding opportunities related to occupations.

Figure 7.2 Example of a retroductive critical realist causation analysis of workers' learning pathways and transitioning experiences, which indicates the intersection of generative mechanisms in open systems at the level of the real, with implications for actual and empirical experiences (which are also real)

Source: Adapted from Mohanoe, 2013, drawing on Bhaskar, 2010

Conclusion

As outlined in the literature review above which reviewed key debates in the learning and work transitioning research literature, our research identified an important gap in this literature which we have sought to address, namely the need for deeper insight into learning and work transitioning experiences framed within a critical vocationalism orientation. We have argued that this is important for transformative green skills research in South Africa, as shown by the two cases above.

However, the more substantive argument of this chapter is that there is need for methodology development for such research, especially methodological approaches that allow for examining learning and work transitioning experiences, both in inter-transitioning (between higher education and work) and intra-transitioning (within workplaces) contexts. We have used two case studies of research undertaken within our green skills research programme to give two examples of how researchers have used meta-theory (critical realism) as well as field-based substantive theory (transitioning theory), human development theory (cf. Bronfenbrenner, 1979) and skills development theory (e.g. theories of articulation and boundary crossing) to design studies that have allowed the researchers to gain in-depth understandings of learning and work transitioning experiences. These transitioning experiences are not, however, limited to narratives of individuals in work and learning transitions (important as these were for the studies); they also offer insight into structural dynamics and causal factors shaping these transitioning experiences. This analysis provides for understanding the structural processes that affect learning and work transitioning pathways such as the way that power relations and workplace relations shape such transitioning opportunities (or not), or how field-based economic histories have shaped and are shaping the demand and need for available training options.

In both the cases shared above, there was need for multi-layered analysis (see also Chapter 6), which included in the first instance historical and document analysis, contextual analysis including field-based observations and in-depth interviewing of carefully selected respondents. It also required careful development of research instruments and questions that allowed for the depth of detail required for understanding the complex and often non-linear nature of learning and work transitioning experiences, with emphasis on both what was present and experienced, and also what was absent. Both the cases above show how important complementing inductive and abductive inferences in the first layer of analysis with retroductive analysis was for identifying those structural processes that could begin to be addressed in order to 'ease' the transitioning processes of individuals, and as shown in case two, not only individuals but also groups of workers who are experiencing similar problems at particular levels of employment.

Acknowledgements

The work of the two Masters in Education case study researchers, Leanne Mckrill and Nthabiseng Mohanoe, is acknowledged as being influential in shaping the arguments in this chapter. The literature review work draws on the PhD study of

Presha Ramsarup. The studies were funded by the South African Qualifications Authority and the Conservation, Arts, Tourism, Heritage and Sports Sector Education and Training Authority. The Department of Science and Technology/ National Research Foundation SARChI Chair in Global Change and Social Learning Systems is also acknowledged for support in supervision of the studies, as well as the writing of this chapter.

References

Archer, M.S. (2000). *Being human: The problem of agency*. New York: Cambridge University Press.

Association for the Development of Education in Africa (ADEA). (2008). Post primary education in Africa. Challenges and approaches for expanding learning opportunities. Synthesis prepared for 2008 ADEA Biennale on education in Africa. Maputo, Mozambique, 5–9 May 2008.

Bhaskar, R. (2010). Contexts of interdisciplinarity. Interdisciplinarity and climate change. In R. Bhaskar, C. Frank, G. Hoyer, P. Naess, & J. Parker (Eds.), *Interdisciplinarity and climate change: Transforming knowledge and practice for our global future*, pp. vii–xiii. London: Routledge.

Bronfenbrenner, U. (1979). *The ecology of human development: Experiments in nature and design*. Cambridge, MA: Harvard University Press.

Bynner, J. (2001). British youth transitions in comparative perspective. *Journal of Youth Studies*, *4*(1), 5–23. doi: 10.1080/13676260120028529

Cohen, L., & Mallon, M. (2001). My brilliant career? Using stories as a methodological tool in career research. *International Studies of Management and Organization*, *31*(3), 48–68.

Department of Health. (1965). Medicines and Substances Control Act of 1965 (Act 101 of 1965). Pretoria: Republic of South Africa.

Eccelstone, K. (2009). Lost and found in transition. Educational implications of concerns about "identity", "agency" and "structure". In J. Field, J. Gallacher, & R. Ingram (Eds.), *Researching transitions in lifelong learning*. pp. 9–27. Hoboken, NJ: Routledge.

Evetts, J. (2009). New professionalism and new public management: Changes, continuities and consequences. *Comparative Sociology*, 8(2), 247–266. doi: 10.1163/156913309X421655

Fenwick, T. (2013). Understanding transitions in professional practice and learning: Towards new questions for research. *Journal of Workplace Learning*, *25*(6), 352–367. doi: 10.1108/JWL-06-2012-0047

Field, J. (2012). Transitions and lifelong learning: Signposts, pathways, road closed? *Life Long Learning in Europe*, *1*, 5–11.

Lindley, D. (2014). *Expansive social learning in wetland management: A case study of the Mondi Wetlands Programme*. Unpublished PhD thesis, Rhodes University, Grahamstown, South Africa.

Livingstone, D. (2004). *The education-jobs gap*. Toronto: Broadview Press.

Lotz-Sisitka, H., & Ramsurup, P. (2011). *Learning pathways, articulation and differentiation: Early conceptual explorations and implications for research design/s*. Unpublished SAQA research report, Rhodes University, Grahamstown, South Africa.

Lotz-Sisitka, H., Ramsarup, P., Olvitt, L., & Mohanoe, N. (2017). Boundary making and boundary crossing in learning pathways access and progression: Voices from the workplace. *SAQA Bulletin*, 137–186.

Mckrill, L. (2015). *Investigation of enabling and constraining factors affecting the supply and demand of specific biodiversity scarce skills to the biodiversity sector.* Unpublished Master's thesis, Rhodes University, Grahamstown, South Africa.

Mohanoe, N. (2013). *Learning pathways of key occupations relevant to sustainable development in Makana municipality*. Unpublished Master's thesis, Rhodes University, Grahamstown, South Africa.

Organisation for Economic Co-operation and Development (OECD). (2008). *OECD employment outlook 2008*. Paris: OECD.

Price, L. (2017). Wellbeing research and policy in the U.K.: Questionable science likely to entrench inequality. *Journal of Critical Realism*, *16*(5), 451–467. doi: 0.1080/14767430.2017.1371985

Raffe, D. (2008). The concept of transition system. *Journal of Education and Work*, *21*(4), 277–296. doi: org/10.1080/13639080802360952

Ramsarup, P. (2017). *A critical realist dialectical understanding of learning pathways associated with two scarce skill environmental occupations within a transitioning systems frame*. Unpublished PhD thesis, Rhodes University, Grahamstown, South Africa.

South African National Biodiversity Institute (SANBI) & The Lewis Foundation. (2010). *A human capital development strategy for the biodiversity sector: 2010–2030*. Pretoria.

Sawchuk, P., & Taylor, A. (2010). Understanding challenging transitions in learning and work. In *Challenging transitions in learning and work*, Vol. 2, pp. 1–25. Leiden, Netherlands: Sense Publishers.

Tikly, L. (2015). What works, for whom, and in what circumstances? Towards a critical realist understanding of learning in international and comparative education. *International Journal of Educational Development*, 40, 237–249. doi: 10.1016/j.ijedudev.2014.11.008

Wildlife Ranching South Africa (WRSA). (2012). Game farmers prevented from using tranquiliser drugs. Retrieved from www.wrsa.co.za/news/item/275-game-farmer sprevented- from-using-tranquilising-drugs

8 Probing the potential of social ecosystemic skills approaches for green skills planning

Perspectives from Expanded Public Works Programme studies

Heila Lotz-Sisitka

Introduction: Research context and problem

South Africa has a long history of exclusion, marginalisation and neglect of worker education and training, and a long history of chronic unemployment, especially for the Black majority. In response, in 1994, the Reconstruction and Development Programme of the first democratic government (African National Congress (ANC), 1994) recommended job creation through a national Public Works Programme (PWP) and an integrated qualifications framework to ensure learning pathways out of poverty for South Africa's people. This led to a National Qualifications Framework and new qualifications and skills development programmes (Atkinson & Ingle, 2016; Giqwa, 2018). These skills system instruments have been used to structure a training component in what has now come to be known as the Expanded Public Works Programme (EPWP) in South Africa, with 'expanded' referring also to expansion beyond traditional PWPs to include skills development. The EPWP is essentially a state-led strategy to alleviate unemployment and poverty, and is widely used in the natural resources management and environmental sector. Stats-SA reported that unemployment levels were at 27.7% in early 2017, meaning that there were 6.2 million people unemployed in a labour force of 22.4 million (Ward, McClean, Kraak, Jenkin, & Mushangai, 2017).

South Africa currently hosts one of the largest public works programmes in the world (Phillips, 2004). Adato and Haddad (2001) indicated that the EPWP in South Africa is amongst the most innovative in the world with multiple objectives which include job creation, poverty reduction, skills development, infrastructure and small medium and micro enterprise development. While this is the case, their full potential, despite focused innovations that emerged from Phase 1, to Phases 2 and 3 of the EPWPs, especially from a skills and learning pathways perspective, remains unfulfilled (Meth, 2011; Atkinson & Ingle, 2016; Burger, 2016; Fourie, 2016; Vaughn, 2016; Ward et al., 2017; Giqwa, 2018). Vaughn (2016) and the Department of Public Works (DPW) (DPW, 2017) indicate that in Phase 3, it is the intention of EPWP programmes to move beyond generic and general lifeskills training towards project based training that is able to support access to work and enterprise development competences.

EPWP skills development programmes are governed by the Quality Council for Trade and Occupations (QCTO). Priority is given to accredited skills programmes in EPWP training, meaning that the Sector Education and Training Authorities have a key role to play in the structuring of the design of the programmes via the unit standards and skills programmes that they accredit. EPWP funds are only disbursed to Sector Education and Training Authority (SETA) accredited training providers resulting in a top-down model that is nationally centralised and managed, at times via provincial structures, and at other times directly from national level to operational levels. Training must be offered during the implementation or at the start of the EPWP programmes. Most of the EPWP training is funded by the Department of Higher Education and Training National Skills Fund, while the DPW (in partnership with the different EPWP programmes) commissions and monitors the training. Learners receive stipends for attending training with higher incentives for attending accredited training, cementing the top-down model. The EPWP directorates in national government departments, such as the Department of Environmental Affairs under the guidance of the Department of Public Works (who oversee all EPWP programmes), produce training guidelines and standard operating procedures for skills programmes, monitor the training and govern the process via tracking accreditation of providers, expenditure, training registers and certification of EPWP beneficiaries. There is little in-depth qualitative analysis of EPWP training programmes (Atkinson & Ingle, 2016; Vaughn, 2016).

Two large scale EPWP programmes operate in the environmental and natural resources management sector and are implemented by the Department of Environmental Affairs. They cover areas of land rehabilitation, water, fire, waste, wildlife and open space management, youth development, rural development and ecological infrastructure, with the unemployed from the communities surrounding the projects being recruited for jobs in the projects (Giqwa, 2018). By 2017, the Environment and Culture sector had supported 1 683 485 job opportunities, with 12.9% or 968 303 of these being green work opportunities (DPW, ILO & EPWP, 2017), with training costing approximately R15 000 per beneficiary (DPW, 2017). Benefits are not only cash effects on consumption smoothing and poverty reduction, but also the productive effects of cash, assets created and skills development (Beazley, Morris & Vitali, 2016). A recent green jobs roadmap (DPW, ILO & EPWP, 2017) projects potential for new green skills areas in sustainable land-based livelihoods, coastal management, waste management, road infrastructure, renewable energy and energy efficiency, water conservation and treatment infrastructure, food security and sustainable farming and gardening, green buildings (public sector and social housing focus), non-motorised transport and eco mobility, disaster reduction measures, environmental education and healthy and sustainable communities.

In 2010 the Department of Environmental Affairs recommended improving the sustainability value of the EPWP skills programmes offered. This catalysed a number of more in-depth research projects into green skills learning potential and learning pathways out of the EPWP programmes covering issues of knowledge flow and the quality of the learning programmes (Giqwa, 2018), occupational research and learning pathways (Burger, 2016; Fourie, 2016; Department of Environmental Affairs [DEA], 2018), and political economy and economic system development for expanding short term job opportunities into viable work streams (Ward et al., 2017). Vaughn (2016) reported that,

> There is insufficient coordination within and between Public Employment Programmes; and no coordination between Public Employment Programmes and related initiatives, such as the development of cooperatives, the provision of adult education and training, and the provision of support for SMMEs. Also, the presentation given at the Green Jobs Summit (April 2014) stated that lessons are not adequately shared across programmes.
>
> (p.59)

Meth (2011), Vaughn (2016) and Atkinson and Ingle (2016) have critiqued the quality and outcomes of EPWP training programmes, and indicate that there is little nuanced understanding of the value of the training. This chapter therefore addresses the need for more in-depth research to develop deeper, more social ecosystemic understandings of green skills in the EPWP sector. This area of research is under-developed with inadequate methodological tools available for undertaking in-depth EPWP skills development studies that offer more qualitative insights into the value and potential of the training and skills development. This has resulted in poorly articulated viable learning pathways and a lack of adequate insights for qualitative improvement of programmes. This chapter addresses these issues, but recognises that a more substantive research process using a social ecosystemic orientation will be required to do full justice to the scope of the challenge. The contribution should therefore be seen as invitational and agenda setting, rather than conclusive.

Theoretical and conceptual tools: Exploring the potential of social ecosystemic approaches for green skills EPWP research

EPWP skills programmes have, to date, attracted mainly national, macro-level quantitative analyses. While useful for monitoring purposes, we extend this research using the concepts and guidance from social ecosystemic research approaches (Hodgson & Spours, 2018) in order to consider a more strongly constituted place-based, connected system of working, living and learning in which EPWP training could potentially play a stronger development role in local communities and contexts. An ecosystemic, place-based approach to skills planning could better benefit EPWP workers who are drawn from local communities and contexts, especially as the DEA (2013) argues that the knowledge and skills transferred and taught to the workers should be relevant to the particular EPWP project and the needs of the communities around the project (Giqwa, 2018). This implies an intention to connect work, living and learning (Thwala, 2011; Hodgson & Spours, 2018). McCord (2002) and Phillips (2004) have indicated that EPWP employment was initially allocated on a household basis with the overall intention of the training component being to enhance capabilities for learning pathways out of poverty. The EPWP favours labour intensive methods and skills development programmes to support these methods, but elementary occupations associated with the jobs are undifferentiated (DEA, 2018).

Dalziel (2011) argued that one of the more useful frameworks for analysing regional systems of skills training is the "skills ecosystem" concept which was developed by authors such as Finegold (1999), Buchanan et al. (2001), Windsor and Alcorso (2008) and Buchanan and Jakubauskas (2010, pp. 44–49). A skills ecosystem, as defined in the majority of the papers that work with the concept, refers to "a self-sustaining network of workforce skills and knowledge in an industry or region" (cf. Windsor & Alcorso, 2008 cited in Dalziel, 2011, p. 3). This body of skills ecosystem research uses the notion of 'ecosystem' to indicate integration, especially integration of perspectives of employers and employees, as these are motivated by market opportunities. Typical applications are found in the finance and technology sectors (e.g. Silicon Valley) with their genesis and application emerging mainly from "the interaction between digital technologies and venture capital" (Hodgson & Spours, 2018).

The models of 'skills ecosystem', developed out of the work of Finegold and Dalziel referred to above, have been critiqued for being elitist given their genesis and application contexts. In response, a 'social ecosystemic' model has been developed by Hodgson and Spours (2016), Hodgson et al. (2017), Hodgson and Spours (2018) and Spours (2019) which "envisages bringing together of a wide range of social partners around the relationship between working, living and learning" favouring and offering a more inclusive model of social collaboration that sees "a particular role for local anchor institutions, such as further education colleges, that can facilitate 'citizen pathways' to new jobs and sustainable living" (Hodgson & Spours, 2018, p. 3). The social ecosystemic model develops a theoretical framework that emerges from Finegold's "High Skills Ecosystem Model, and a spatial adaptation of Uri Bronfenbrenners's Human Ecological Systems (Bronfenbrenner, 1979), together with the addition of a third element – concepts of mediators and ecosystemic leadership" (Hodgson & Spours, 2018, p. 4). Theorists of this model argue that there is need to develop experimental platforms "for collaborative working between important social partners from which lessons can be learned in building the larger long-term social formation" (ibid.), indicating that development of such approaches requires collaborative, place-based engagement with multiple actors over time. The attraction of considering this theoretical framework for green skills EPWP programme research and planning lies in the argument cited below, which aligns well with the chapters in this book:

> human ecosystem thinking has the potential to help us to understand how different social actors can come together to create dynamic clusters of economic, social and sustainable activity. Seen in this way, human ecosystem thinking could also be viewed as leading to a different model of the state and economy beyond the increasingly worn-out neoliberal binaries of

> marketization and managerial state, and as a movement towards a more collaborative, devolved, networked and civil society concept of governance and leadership. At a broader ethico-political level still, human ecosystem thinking is also underpinned by a growing awareness of the enormous challenge of sustainability and the crisis of the relationship between humanity and the planet and, in this sense, it can be seen as a representation of the current zeitgeist.
>
> (Hodgson & Spours, 2018, p. 6)

Cronin (2016, p. 24) argued this in a developmental state frame, indicating that in such a model it is necessary to nurture relationships between state and citizens in which "the notion of top down delivery is broken down, and belief in the power of popular agency and participation is fostered in its place". He argued that the impact of EPWP programmes must be improved at the household and community level with a more locally centric approach to EPWP programme implementation. Importantly, and aligned with this sentiment, the social ecosytemic model requires giving attention to relationships with a '*facilitating national state*' (e.g. the EPWP planning and monitoring units of the DPW and DEA), a devolved local state (e.g. provincial and local government or parastatal implementation partners in EPWP programmes via provincial development plans (PDPs) and local government integrated development planning (IDPs)), via development of public/private networks with a strong role for anchor institutions (e.g. technical and vocational education and training (TVET) colleges, or local economic development (LED) offices), as well as '*horizontal place-shaping*' relations involving a "strategic mix of public, private and third sector inputs through networks" with community, work and democratic participation being foregrounded (Hodgson & Spours, 2018, p. 12). Importantly for green skills research, the skills ecosystemic concept focuses on "understanding the wider context or setting in which skills are developed and used as part of moving the policy and practice debate beyond the prevailing model of 'skills supply'" (ibid., p. 12; cf. also Chapters 1, 2, 3 and 11 in this volume),which currently dominates the approach to conceptualising, planning and monitoring EPWP training programmes. Spours (2019) argued that social ecosystemic skills research should involve what he calls "45-degree activity" at the interface of 'facilitating verticalities' and 'collaborative horizontalities' (see Figure 8.1). While not theorised as a critical realist lamination, the model reflects a multi-levelled or ontologically laminated system perspective in which emergence occurs variously within and between the laminations (cf. Chapter 12).

In the Spours (2019) model, developed in the context of skills development in an urban environment (East London in the UK), he foregrounds social ecosystemic leadership, education practice/systems and socialised digital technologies as being important to '45-degree activity' with this referring to the connecting skills development activity that mediates between 'connecting horizontalities' and 'facilitating verticalities'. He argued further that social-ecosystemic skills planning leadership should be place-based and should involve capability to define the problem terrain, and the goal of inclusive sustainable social, economic and educational development. In our green skills research (see below) we would also include a focus on the actual ecological context of the place-based skills ecosystemic context, as well as an in-depth sociological understanding of the histories of social marginalisation and ecological degradation, their emergence and contemporary cultural historical configurations (Fourie, 2016; Giqwa, 2018; Pesanayi, 2019).

Spours (2019) argued that social-ecosystemic skills leadership also needs to focus on "skills escalators as 'citizen pathways'" in order to facilitate progression within education, as well as access to employment and progression within work, offering a broader framing of occupational progression as theorised by Ramsarup (2017) (cf. Chapter 5). In the EPWP context this would involve being able to articulate and/or develop such pathways into education, employment and work (ibid.). He argued further that this requires giving attention to the linking of low, intermediate and high skills within a particular context (cf. Ramsarup, 2016), co-production and experimentation involving multiple partners that can contribute to progressions and the emergence of higher levels of skills that are appropriate to the developmental context (cf. Chapter 10, Lotz-Sisitka et al., 2016; Pesanayi, 2019). In the EPWP context, there is need to also develop knowledge of the social-ecological and social-economic development interventions that are being serviced by EPWP 'jobs' for their current and future employment creation potentials (Burger, 2016; Ward et al., 2017) as this area is still under-developed in South African research, i.e. there is not enough to focus on skills development only, there is also need to focus on developmental trajectories and how they create demand for green skills (cf. Chapter 3). Kraak (2017, cited in Ward et al., 2017) also argued that there is a need to focus on the efficacy of local actors and institutions within a skills ecosystem, especially their roles in providing complementary support services relied on by the other players in the ecosystem.

The Hodgson and Spours (2018) social ecosystemic model also highlights an important role for learning institutions such as technical education and training and further education and training colleges in cementing the potential skills development pathways into local economies over time, with a role of not only providing skills, but also operating as developmental 'hubs' in partnership with local development and business entities. In a South African context, this would require a combination of local economic development (LED) structures working closely with relevant learning institutions, such as TVET colleges, agricultural training institutes or community colleges, and where these are absent other training institutions such as non-governmental organisations (NGOs), faith based organisations, etc. especially in rural areas where formal learning infrastructure is often sparse, or limited to schools (cf. Lotz-Sisitka et al. 2016; Pesanayi, 2019: for an emerging local example in the agricultural learning system context, cf. Chapter 10). In the EPWP context, there would also be a need to integrate current modes of 'private provision', which are often mobile units, into more situated and longer lasting community college programme offerings. This has implications for the way in which accreditation systems provide accreditation to skills providers and how skills funding is disbursed by the state via the EPWP programmes to training providers (i.e. the 'facilitating verticalities' may also need to change their practices and rules). Spours also mentioned the potential role of 'socialised digital technologies' for 45-degree activity, an area that is highly under-developed for its potential in EPWP skills development programmes in South Africa, despite these technologies having particular affordances that can also enhance social learning (Durr & Lotz-Sisitka, 2019).

Exploring some '45-degree activity' EPWP green skills research approaches

As indicated in Figure 8.1, there is need to develop green skills research methodology for probing 45-degree activity within a social ecosystemic orientation that seeks to better connect work, living and learning in the EPWP context. Here, I briefly describe a few different *types of studies* that each offer research tools and orientations that can be further developed within a more substantive framing of green skills social ecosystemic research for sustainable development. This illuminates potential methodological lenses on green skills research in the 45-degree activity space illustrated via the adapted Figure 8.1. These offer an early 'experimental platform' for re-configuring green skills EPWP research. All the studies are situated in various ecological settings (e.g. rehabilitation landscapes, rural urban waste management, alien invasive clearance, etc.) and all

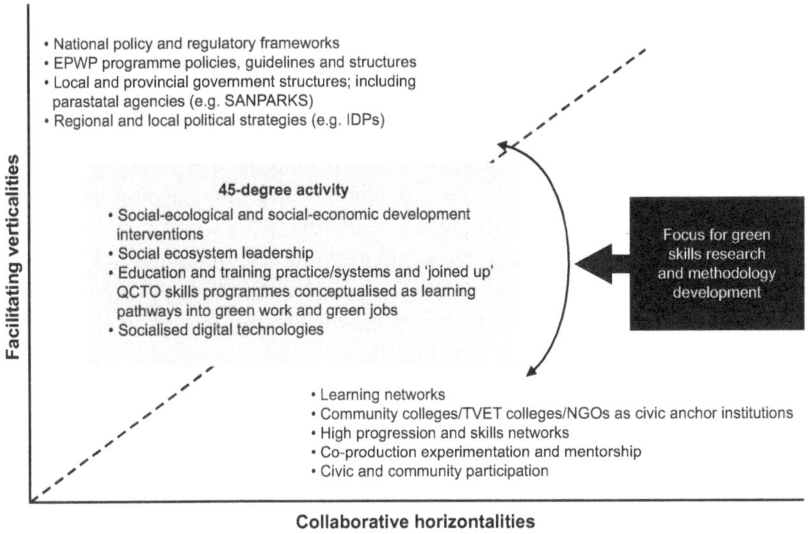

Figure 8.1 Focus on '45-degree activity' in social ecosystems research in green skills research and methodology development

Source: Adapted from Spours, 2019

seek in-depth insights into skills development in the contexts where the EPWP programmes are operating. In all cases, the EPWP training on offer lacks adequate quality and learning pathway options (Burger, 2016; Fourie, 2016; Giqwa, 2018). Often occupations related to the training and the short term jobs either do not exist or are poorly articulated in the occupational system (Fourie, 2016, DEA, 2018) and are therefore misaligned with the 'facilitating verticalities', or are beyond the reach of people who are seeking to move out of poverty, i.e. lack adequate institutional support, showing misalignment with 'collaborative horizontalities'. The studies therefore not only illuminate the dynamics of the '45-degree activity' space that Spours refers to, but also illuminate absences and weaknesses in the 'facilitating verticalities' and 'collaborating horizontalities' within a wider laminated constellational reality (cf. Chapter 12).

The methodologies are not discussed in great detail, but rather pointed to for their potential offering to the development of social ecosystemic approaches to EPWP skills development. Readers are referred to the actual studies for more depth insight into the methodologies (meaning combination of theory and method).

- **Study 1: Knowledge recontextualisation and legitimation research** (Giqwa, 2018): An EPWP green skills study by Giqwa (2018) offers methodological tools for probing the quality of education and training system interventions in EPWP skills programmes. Using Bernstein's (1990) pedagogic device and recontextualisation theory, as well as its further development via Maton's (2014) legitimation code theory, Giqwa (2018) investigated the recontextualisation and circulation of waste management knowledge in EPWP programmes. She used qualitatively rich analysis to track how waste management knowledge was produced by scientists and policy makers, and then recontextualised from policy level (via skills programme qualification unit standards governed by the QCTO) to training provider level (accredited training programmes and materials) to worker training programmes (observations of training offered) and into ongoing workplace learning interactions. The study was undertaken in rural areas in South Africa, given the EPWP aim to contribute to rural development. The study found that the knowledge structure that dominated the programmes had a 'strong epistemic code' meaning it was technically inscribed. This knowledge code clashed with the dominant knowledge code of workers in the EPWP whose knowledge was characterised by a strong 'social code' meaning it was more contextualised and locally inscribed. While this is not necessarily a problem and one would anticipate that training programmes would seek to expand workers' existing scope of knowledge, the study also found significant pedagogical problems that emerged around this knowledge code clash as the training providers were not able to mediate the knowledge effectively. This meant that workers learned very little in the training programmes which were dominated by technical concepts that did not connect with workers' knowledge base. The study showed that a range of practical activities helped to develop understanding of the technical knowledge. The study also found that the knowledge on offer in the unit standards and therefore also the training programmes were narrowly constituted around general technical issues, and were not related to local economic and other opportunities. For example, workers were taught to sort waste but not to relate this to management of local landfills or enterprise development opportunities that were possible through other applications of waste in the local development value chain. This methodological approach allowed for in-depth epistemic understanding of workers and providers and how their interactions either facilitate or impede learning in the social ecosystemic context, showing the need especially for providers to more adequately contextualise the programmes on offer. Vertical facilitations would require more flexible and responsive qualifications and a different logic for quality assurance and mediation of the training. This is because the study also showed anomalies of Level 2 programmes being implemented in a highly prescriptive, top-down fashion, even when many of the workers already had Level 4 qualifications (i.e. programmes were not adapted to the learners' prior knowledge and/or contemporary learning needs or potential or the settings in which they were being implemented).

- **Study 2: Capabilities research underlaboured by social realism** (Fourie, 2016): An EPWP green skills study by Fourie (2016) offers methodological tools for probing how environmental learning programmes respond to Working for Water (WfW) learners' capabilities and the enabling and constraining factors that shape these, and to the demands of the WfW training setting in an emerging Green Economy. Using Sen's (1999) concepts of capabilities (people's valued beings and doings) and conversion factors, the study investigated the environmental, personal and social conversion factors that people could draw on to make use of the training offered by the EPWP programme to translate the benefits of the training into accessing more sustainable green jobs in an emerging green economy in an urban context. It further probed the structural, cultural and personal emergent properties (Archer, 1995) that enabled or constrained training participants (workers and contractors, i.e. at two levels) from benefitting from the training programmes offered, providing good insight into personal, regional, ecological and other environmental dynamics that intersect in learning pathways emergence in contexts of poverty. With this in-depth insight into the lives of the learners and the context in which they find themselves, the study probed the academic and practical suitability of the accredited Environmental Literacy training programme being offered by the EPWP (meant to empower learners to get jobs in the emerging green economy) and how it responded to (or not) the education levels and experiences of contractors and worker beneficiaries in the EPWP programme.

  The study found that while the training programme was able to respond to the learners' capabilities and interests to some extent, there is a lack of systemic information as to where the skills can be applied in elementary green occupations, in part because South Africa has over-emphasised high skills for green economy development (cf. Chapter 6; Ramsarup, 2017). The study made recommendations to link training in EPWPs to career guidance and identified the need for more in-depth research into elementary level green occupations and associated learning pathways (Fourie, 2016). This methodological approach allows for an in-depth sociological understanding of workers and providers and how their resources are (or are not) translated into functioning and capabilities or valued beings and doings and viable learning pathways into local green economies that can overcome the historical imprint of a low skills regime (McGrath, Badroodien, Kraak, & Unwin, 2004). This shows the need to link training provision with skills demand analysis in regional and local economic contexts for elementary occupation development.

- **Study 3: Learning and work transitioning and competence research with a learning pathways framework from 'job opportunity' to Small and Medium Enterprise contractor** (Burger, 2016): An EPWP green skills study situated in an urban local municipality undertaken by Burger (2016) traced the learning and work transitioning system support and competences required for translating experience in a Working for Ecosystems 'job opportunity' into viable jobs in the form of Small and Medium Enterprise Contractors operating within the Working for Ecosystems green economy. The study examined the emergence of these workplace-based learning pathways in practice, the skills development in terms of the combination of training and workplace-based learning and how this was oriented towards capacity building for entrepreneurship and social-ecological adaptive management (Folke, Hahn, Olsson, & Norberg, 2005). It also examined the EPWP skills programmes as an opportunity for acquisition of knowledge, competence and new skills development in relation to Small and Medium Enterprise (SMME) practices (i.e. how these programmes are structured and which policies are in place that shape them in particular ways). The study points to the complex intersection of training and practice, prior knowledge, socialisation in the workplace, a growing passion for a practice over time, ongoing support involving formal and informal learning processes combined with infrastructure and other forms of support. The study also points to the significance of tapping into foundations of indigenous plant knowledges for mobilising interest in the area of work and practice (cf. Giqwa's study above on workers' knowledge codes) being central to pedagogical and practice efficacy. The study also shows the shift in competences required for transitioning from 'job opportunity', which involves managing alien invasive plants, to contractor, which involves system and entrepreneurial practice management (i.e. two different types of work). Burger's study found an overemphasis on training for the former in EPWP programmes, with almost no formal training accreditation on offer for the latter. In this context it was the range of other actors (especially NGOs) who took social ecosystemic leadership to support the learning and work transitioning process. The transitioning process could, however, potentially have been achieved in a much shorter time (transitioning time was on average eight years per worker-in-transition) if stronger forms of national facilitatory support (e.g. in the form of an accredited enterprise development course for EPWP workers) had been available to combine with the horizontal collectivities being developed via the municipality and local NGO system of support for these workers. This study points to the importance of adopting a social ecosystemic approach to skills transitioning for viable green work in the EPWP programmes as singular or narrowly constituted top-down training programmes are inadequate to support the complex process of transitioning (cf. also Chapters 6 and 7). Burger's (2016) study also points to the need to establish stronger political economies for work out of EPWP programmes in response to the difficulties experienced by contractors via what they termed the problem of 'contracts in break form'.

The studies above are complemented by an emerging body of studies in our research programme that focus on boundary crossing expansive learning research (drawing on the work of Engeström (2016) as well as Engeström and Sannino (2010). This approach to research brings different activity systems in the regional skills development context together to co-engage contradictions around their shared object of contributing to the intended outcomes of the EPWP programmes contextualised in local social-ecological development contexts (Pesanayi, 2019; Thifhulufwelwi, 2019). Via formative intervention research using a method of Boundary Crossing Change Laboratory Workshops (Engeström & Sannino, 2010, cf. Chapter 10), a diversity of actors in the setting deliberated ways of providing a wider range of learning options and learning pathways within a learning network approach. The learning network approach diversifies the type of training beyond accredited training, complementing available accredited training with community engaged learning programmes, media use and use of digital technologies in order to provide for a richer, social ecosystemic approach to learning that can extend training interventions such as those provided by the EPWP programmes in local social-ecological environments (Lotz-Sisitka et al., 2016; Pesanayi, 2019). In these models EPWP training becomes part of a wider system of interrelated education and learning interventions and processes that are joined up around shared activity on a landscape of practice offering wider views of learning pathways and potentials for a wider range of learning processes beyond the 'learning on or for the job' only. The expansive learning methodology used by these studies, which link diverse activity system and communities of practice on a landscape of practice, offers ways of strengthening the 'horizontal collectivities' identified by Spours (2019), while also challenging disjunctures in the 'national facilitatory' structures (Thifhulufwelwi, 2019). Cultural historical activity theory research and expansive learning approaches allow for in-depth understanding of how boundary crossing learning can be established in landscapes or regions of inter-connected practice, linking supply, transitioning and demand systems in a social ecosystemic model for skills development for sustainable development.

The above studies focus mainly on the skills development processes and, as noted by Kraak (2017, cited in Ward et al., 2017), the other elements of the skills ecosystemic need to be critically and carefully considered, hence there is also need to engage in substantive value chain and green economy development system studies and evaluations to articulate viable learning pathways into green jobs in the EPWP space as there are also many assumptions about 'work' and the sustainability of work and job opportunities in the EPWP programmes (DPW/ILO/EPWP, 2017; Ward et al., 2017). As yet, the EPWP programmes remain untapped for their potential for value addition into small and medium enterprise development, although this is becoming a focus for the programmes (Ward et al., 2017), but it will require a re-orientation of the 'short termism' which underpins the notion of 'job opportunity' in EPWPs, in order to explore the potential for local economic development opportunities and job creation potential to emerge (Burger, 2016; Ward et al., 2017). These studies are showing the need for more in-depth understanding of *the processes of interconnection in social ecosystemic skills development environments*. Ward et al. (2017) stated that the actual information on skills development within this relationship remains poor. These studies also argue for locating social ecosystemic green skills research approaches (focused on the 45-degree activities) within wider systemic perspectives, especially wider political economies shaping sustainable development.

Conclusion

As shown above, there is much to be done to uncover the complex dynamics of a social ecosystemic approach to skills development, especially if it is to be place-based and if it is to offer realistic and meaningful learning, work and living options to people who are currently trapped in poverty. In the context of green skills, a massive national resource such as the EPWP programme has the potential to make a huge impact on local sustainable development, if the programme itself can adapt and develop more regionally constituted orientations that can offer substantial inputs into regional development initiatives. For this to happen, a different development logic is required, and a greater balance of facilitating verticalities (not top-down vertical imposition) and horizontal collectivities is needed, with more substantively constituted approaches to skills development offering '45-degree activities' to connect these levels iteratively and qualitatively. This chapter has shown that the social ecosystemic approach can produce a stronger synergy between education, training, workforce development and sustainable livelihoods development, if attention is also given wider to economic, political and ecological forces.

References

Adato, M., & Haddad, L. (2001). *How efficiently do public works programmes transfer benefits to the poor?* Discussion Paper 108. Washington, DC: International Food Policy Research Institute.

African National Congress (ANC). (1994). *The reconstruction and development programme: A policy framework*. Johannesburg: Umanyano Publications.

Archer, M.S. (1995). *Realist social theory: The morphogenetic approach*. Cambridge: Cambridge University Press.

Atkinson, D., & Ingle, M. (2016). Chapter 9: A review of training in the EPWP. In Vaughn, A. (Ed.) *Our expanded public works programme: Reflections on over a decade*, pp. 135–150. Johannesburg: South African Cities Network.

Beazley, R., Morris, R., & Vitali, A. (2016). Designing public works programmes and cash transfers to promote growth and mitigate violence. EPS PEAKS, UK Retrieved from www.gov.uk/dfid-research-outputs/designing-public-works-programmes-and-cash-transfers-to-promote-growth-and-mitigate-violence

Bernstein, B. (1990). *Class, codes and control, Volume 4: The structuring of the pedagogic discourse*. London: Routledge.

Bronfenbrenner, U. (1979). *The ecology of human development: Experiments by nature and design*. Cambridge, MA: Harvard University Press.

Buchanan, J., Schofield, K., Briggs, C., Considine, G., Hager, P., Hawke, G., Kitay, J., Meagher, G., Macintyre, J., Mounier, A., & Ryan, S. (2001). *Beyond flexibility: Skills and work in the future*. Sydney: NSW Board of Vocational Education and Training. Retrieved from www.bvet.nsw.gov.au/pdf/beyondflex.pdf

Buchanan, J., & Jakubauskas, M. (2010). Chapter 3: The political economy of work and skill in Australia: Insights from recent applied research. In J. Bryson (Ed.) *Beyond skill: Institutions, organisations and human capability*, pp. 32–57. London: Palgrave Macmillan.

Burger, M. (2016). *Working for ecosystems: An account of how pathways of learning lead to SMME development in the green economy*. Unpublished Master's thesis, Rhodes University, Rhodes University, Grahamstown, South Africa.

Cronin, J. (2016). Chapter 2: Public employment programmes and the future of work. In Vaughn, A. (Ed.) *Our expanded public works programme: Reflections on over a decade*, pp. 21–34. Johannesburg: South African Cities Network.

Dalziel, P. (2011). Leveraging training and skills development in SMEs: A regional skills ecosystem case study. New Zealand: Lincoln University, AERU. Retrieved from http://researcharchive.lincoln.ac.nz/bitstream/handle/10182/4439/dalziel_sme.pdf?sequence=1

Department of Environmental Affairs (DEA). (2010). *Environmental Sector Skills Plan of South Africa. A systems approach to human capacity development and sector skills planning*. Pretoria: Department of Environmental Affairs.

Department of Environmental Affairs (DEA). (2013). *Environmental Protection and Infrastructure Programme Framework*. Pretoria: Department of Environmental Affairs.

Department of Environmental Affairs (DEA). (2018). *Research on Educational and Occupational Pathways for the Expanded Public Works Programme Participants of the Environmental Programmes*. Report prepared by Green Skills Academy for DEA. Pretoria: Department of Environmental Affairs.

Department of Public Works (DPW). (2017). *Expanded Public Works Programme (EPWP) Progress Report*, with Deputy Minister. Public Works and Infrastructure. 8 February 2017. Parliamentary Monitoring Group. Retrieved from https://pmg.org.za/committee-meeting/24053

Department of Public Works (DPW), International Labour Organisation (ILO), & Expanded Public Works Programme (EPWP). (2017). *Developing a green jobs roadmap for the Expanded Public Works Programme (EPWP)*. Green Jobs Roundtable presentation, 18 July 2017, Council for Scientific and Industrial Research (CSIR) International Convention Centre, Pretoria, South Africa.

Durr, S.J., & Lotz-Sisitka, H.B. (2019). *Exploring the social learning value enabled by affordances of the Food for US mobile application: A South African food redistribution case study*. 15th International Conference on Mobile Learning 2019. Utrecht, Netherlands. 11–13 April 2019.

Engeström, Y. (2016). *Studies in expansive learning. Learning what is not yet there*. New York: Cambridge University Press.

Engeström, Y., & Sannino, A. (2010). Studies of expansive learning: Foundations, findings and future challenges. *Educational Research Review*, 5(1), 1–24.

Finegold, D. (1999). Creating self-sustaining, high-skill ecosystems. *Oxford Review of Economic Policy*, *15*(1), 60–81.

Folke, C., Hahn, T., Olsson, P., & Norberg, J. (2005). Adaptive governance of social-ecological systems. *Annual Review of Environment and Resources*, 30(1), 441–473. doi: 10.1146/annurev.energy.30.050504.144511

Fourie, K. (2016). *A critical review of the response of the Environmental Literacy Skills Programme to learner capabilities, and to the demands of the Working for Water training setting in an emerging green economy*. Unpublished Master's in Education thesis, Rhodes University, Grahamstown, South Africa.

Giqwa, N. (2018). *Waste management knowledge, its production, recontextualisation and circulation in Expanded Public Works Programme (EPWP) training programmes*. Unpublished PhD thesis, Rhodes University, Grahamstown, South Africa.

Hodgson, A., & Spours, K. (2016). *The evolution of social ecosystem thinking: Its relevance for education, economic development and localities: A stimulus paper*. London: Centre for Post-14 Education and Work, UCL Institute of Education.

Hodgson, A., & Spours, K. (2018). *A social ecosystem model: Conceptualising and connecting working, living and learning in London's New East*. The East London Vocational Education and Training: Innovation through Partnership Programme (ELVET). ELVET Research Briefing No. 3, March 2018, London.

Hodgson, A., & Spours, K. with Jeanes, J., Smith, D., Vine-Morris, M., Vihriala, R., Bollam, J. Coles, L., Harris M., & Kazempour, T. (2017). Education, skills and employment in East London: An ecosystem analysis. ELVET Programme Research Briefing 1. London: Centre for Post-14 Education and Work, UCL Institute of Education.

Lotz-Sisitka, H., Pesanayi, T., Weaver, K., Lupele, C., Sisitka, L., O'Donoghue, R., Denison, J., & Phillips, K. (2016). Water use and food security: Knowledge dissemination and use in agricultural colleges and local learning networks for home food gardening and smallholder agriculture. Volume 1: Research and Development Report. WRC Research Report No. 2277/1/16. Pretoria: Water Research Commission.

Maton, K. (2014). *Knowledge and knowers: Towards a realist sociology of education*. London: Routledge.

McCord, A. (2002). *Public works as a response to labour market failure in South Africa*. CSSR Working Paper 19. November. Retrieved from www.saldru.uct.ac.za/pwp/papers/PWP%20as%20a%20Response%20to%20Labour%20Market%20Failure%20WP19.pdf

McGrath, S., Badroodien, A., Kraak, A., & Unwin, L. (Eds.) (2004). *Shifting understandings of skills in South Africa: Overcoming the historical imprint of a low skills regime*. Cape Town: HSRC Press.

Meth C. (2011). *Employer of last resort? South Africa's expanded public works programme*. Southern Africa Labour and Development Research Unit Working Paper Number 58. Cape Town: SALDRU, University of Cape Town.

Pesanayi, V.T. (2019). *Boundary-crossing learning in agricultural learning systems: Formative interventions for water and seed provision in southern Africa.* Unpublished PhD thesis, Rhodes University, Grahamstown, South Africa.

Phillips, S. (2004). *The EPWP – overcoming underdevelopment in South Africa's second economy.* Paper presented at the United Nations Development Programme (UNDP), Human Sciences Research Council (HSRC) and Development Bank of Southern Africa (DBSA) Conference on Overcoming Under-development in South Africa's Second Economy, 29 October 2004.

Ramsarup, P. (2016). Absenting the absence of parallel learning pathways for intermediate skills: The 'missing middle' in the environmental sector in South Africa. In Price, L., & Lotz-Sisitka, H.B. (Eds.). *Critical realism, environmental learning and social-ecological change*, pp. 116–136. London: Routledge.

Ramsarup. P. (2017). *A critical realist dialectical understanding of learning pathways associated with two scarce skill environmental occupations within a transitioning systems frame.* Unpublished PhD thesis, Rhodes University, Grahamstown, South Africa.

Sen, A. (1999). *Development as freedom.* New York: Alfred A. Knopf.

Spours, K. (2019). Education/Employer partnership working and place-based skills development. Presentation at the Institute of Education, University College London.

Thifhulufwelwi, R. (2019). *How social learning mediation can strengthen resilience building in the context of natural resource management in the Blyde catchment area.* Unpublished PhD research upgrade proposal. Rhodes University, Grahamstown, South Africa.

Thwala, W.D. (2011). Public works programmes as a tool to address unemployment and skills shortages among the youth in South Africa. *African Journal of Business Management*, 5(15), 6011–6020.

Vaughn, A. (2016). Chapter 4: A review of training in the EPWP. Lessons learned from Phase 1 and 2 towards improving Phase 3. In Vaughn, A. (Ed.) *Our Expanded Public Works Programme: Reflections on over a decade*, pp. 53–68. Johannesburg: South African Cities Network.

Ward, M., McClean, D., Kraak, A., Jenkin, N., & Mushangai, D. (2017). *The eco-furniture programme. An evaluative review.* Unpublished report, Pretoria: Department of Environmental Affairs.

Windsor, K., & Alcorso, C. (2008). Skills in context: A guide to the skill ecosystem approach to workforce development, Sydney: New South Wales Department of Education and Training. Retrieved from www.training.nsw.gov.au/businesses/training_options/managing_workforce/skill_ecosystem.html

9 Framing learning needs assessments for sustainability policy practices

Eureta Rosenberg

Learning needs assessments in context: Driving national transitions towards inclusive green economies, through skills development

The Green Economy Learning Assessment for South Africa (GELA-SA) was commissioned in 2015 by PAGE, an international Partnership for Action on the Green Economy. As noted in Chapter 2, the green economy has been a key concept in international and national initiatives to shift governments, industries and society as a whole along the transition to a more environmentally sustainable and socially just future.

The PAGE partners include the United Nations (UN) Institute for Training and Research (UNITAR), the UN International Development Organisation (UNIDO), the UN Environment Programme (UNEP), the UN Development Programme (UNDP) and the International Labor Organisation (ILO) as well as the South African government departments of Environmental Affairs (DEA), Science and Technology, Economic Development and Higher Education, among others. These entities have a strong focus on *policy* as a driver of the green economy, and on skills development as a driver in turn of policy development, implementation and review. PAGE was interested in the capacity of practitioners working in policy contexts, positioned as 'change agents' working in national or regional government or similar contexts, either developing the green economy policy or ushering green economy policy into institutional practice. To strengthen the capacity of existing green economy 'champions', and to produce new champions, PAGE partners wanted to invest in new courses or the expansion of suitable existing courses. In order to evaluate and design courses for 'green economy champions', they needed to understand the learning needs of these policy practitioners better. For this reason, green economy learning assessments were commissioned, the first being an assessment for Ghana, produced in 2015 (www.greengrowthknowledge.org/resource).

The process of conducting the Green Economy Learning Assessment for South Africa has been detailed in the report (PAGE, DEA, UNITAR, & Rhodes University, 2016) and discussed in the journal *Higher Education, Skills and Work Based Learning* (Rosenberg, Lotz-Sisitka, & Ramsarup, 2018).

GELA-SA: Measures taken to ensure credible findings

In reflecting on how the GELA-SA was undertaken, I highlight measures taken to achieve credible findings, as it speaks to the usability of green skills research in relation to its framing. The design of the study was itself a consultative process, involving most members of the PAGE steering committee. Given the partners' interest in a broad identification of learning needs, the methodology differs from a skills audit, which more typically would focus on an organisation to identify both existing skills and actual or predicted future skills needs (often described as 'skills gaps'). It is also useful to consider the differences between an *assessment* (as required by PAGE) and a systematic *review*.

These differences are usefully outlined in the 2000 Millennium Ecosystem Assessment (www.milleniumecosystemassessment.org) which noted that a systematic review generally has a 'scientific' knowledge creation purpose, is typically of longer duration, and focused on comprehensiveness of coverage, depth and theory building. An assessment, on the other hand, has a more immediate applied purpose, typically of a political-institutional nature, i.e. to guide government, policy makers, funders and other institutions. It is typically of shorter duration. However, (learning needs) assessments still need to be credible. Precisely because they are intended to inform decisions about investments of time and funding and policy decisions, they need to be trustworthy, free of vested interests and blind spots.

In the case of the GELA-SA, we aimed to achieve credible findings through a careful mix of methods and processes involving:

- Periodic guiding inputs (through meetings and draft report reviews) from a steering committee that represented PAGE member organisations
- A brief review of all relevant policy documents, and meta-reviews of policy reviews, to scope the assessment
- A review shaping the formulation of a conceptual framework for the study (see below)
- A consultative workshop to further scope the focus and conceptual framing of the assessment, and to obtain member feedback on the chosen approach
- Two online surveys among key stakeholders – green economy 'champions' and training providers – on the required competencies
- Case studies to complement the broad sweep of survey findings, with more in-depth explanatory insights; these were based on:
    - Interviews with a small group of diverse key informants, 'champions' and training providers in chosen sub-sectors
    - Document analysis, perusing project reports and other documents pertaining to the four case study contexts
- A final consultative workshop with key informants and PAGE partners to discuss the findings and recommendations and refine the latter.

This process produced a credible report on the competencies required by practitioners driving green economy initiatives in policy contexts, and the types of learning opportunities that seem to best support the development of these competencies, among both students and professionals already in the workplace. It made recommendations that PAGE shared and acted upon. A new online short course was developed and the researchers were asked to share their findings at national and international forums. However, it was also clear that the recommendations were interpreted by stakeholders in diverse ways and would have benefitted from more critical clarification on the framing of sustainability related learning needs. This is the focus of the remainder of the chapter, in which it may become evident that an online short course intervention could be a very limited and even misdirected investment in skills for green economy initiatives in policy contexts.

To understand this statement, one has to understand the context and nature of sustainability policy practitioners' work. Some examples are shared in this chapter (and elsewhere in this volume). One also has to unpack the notion of competence and how it has been framed in the GELA-SA and the literature on which it drew. This is done in the next section.

Conceptually framing learning needs for assessment

For a conceptual framing of the learning needs of current or future policy practitioners working with or towards a green economy focus, we started with the concept of competence. (Note: 'competence' is equated here with the term 'skill'. 'Competence' is used in this chapter not as a departure from the use of 'skill' in relation to learning needs in the rest of the book, but simply to be consistent with the terminology in the GELA-SA, which precedes the book.)

We used the term competence in the GELA-SA to denote a complex of knowledge, ethical and practical application abilities among, in this case, graduate level professionals. Because of our focus on graduate level professionals in what could be constituted as a broader sustainability context, we drew on and adapted a definition of competence used by Wiek, Withycombe and Redman (2011) in their meta-level study of the learning needs that are particular to the field of sustainability science, to mean "a functionally linked *complex*" of knowledge, abilities and dispositions that "enable successful task performance and problem solving". That notion of competence was useful in that it alludes to the intertwined technical, knowledge-based and values-based or normative nature of the abilities that people use to undertake work in the typically demanding contexts of complex sustainability practices, such as introducing a new green procurement policy, or implementing a pioneering, transformative water law. (Note: Sustainability science is not here equated with sustainability policy practices, that is to say, we do not assume that policy practitioners are necessarily scientists, although they may be. Rather, we draw on the competency framing that Wiek and colleagues used, because it seems generically applicable for this cross-disciplinary field and level of work. Furthermore, we substitute 'sustainability policy-practice' for 'green economy policy-practice' whenever a wider range of policy-practices seems relevant.)

However, the concept of competence can be used in problematic ways. This is discussed next in order to demonstrate the importance of conceptual scrutiny and coherence when designing green skills studies, in particular, learning needs assessments (and related enquiries, such as training needs or skills audits).

The notion of 'competence' is open to interpretation. One of the more problematic approaches is to interpret it within a neo-liberal labour market orientation (Gamble, 2018), which positions it within a presumed one-to-one relationship between education (or training) and work. This is a reductionist framing, as it reduces the often complex undertaking of work (demonstrated in the studies of greener value chains in mining, agriculture and paint) to a series of easily identifiable and discrete tasks. Within a neo-liberal framing, competencies are described in terms of "satisfactory work performance" or "effective performance" (Wheelahan, 2007; Gamble, 2018). This performativity discourse implies that work standards are in place for a particular task, and that the task and a desired outcome can be tightly circumscribed beforehand. This may be applicable to a task such as "draw up a legal contract" but not to the more open-ended tasks, such as setting up South Africa's first public-private partnership for the procurement of renewable energy. As shown below, such a complex undertaking required the development of new systems and processes, which could draw on international best practice, but also had to be informed by successive rounds of action, reflection and re-design, that is, such work is to a large extent emergent and cannot be approached and adjudged (only) in terms of "effective performance".

Breaking work up into tasks, and breaking tasks up further into knowledge, behavioural and attitudinal components, is nonetheless a common practice that allows education providers and human resource managers to design training modules and 'key performance indicators', and skills researchers to catalogue learning needs. However, lists of desired competencies tend to fail to adequately reflect and encompass the integrated, emergent and responsive nature of (green) work in complex contexts.

Another critique of the way in which competencies is sometimes used for skills research and planning is that it risks undervaluing the role of foundational disciplinary knowledge and generic analytic abilities (Wheelahan, 2007) on which practitioners draw to engage with tasks that are novel, emergent or deviating from what was known or done before, under the influence of changing contextual drivers. This point is important, perhaps especially so in relation to undergraduate training, when much foundational knowledge is still being acquired and developed.

Lotz-Sisitka (2018), reflecting on the notion of 'change agents', also pointed to an individualistic fallacy in the learning needs assessment discourse, which tends to assume that people apply their skills and knowledge as isolated individuals in decontextualised situations. Instead, she argues, competencies to respond to complex sustainability challenges and opportunities are distributed among groups of people, and never reside only in an individual. Furthermore, they are brought forth within these collectives in relation to real world challenges, that is, situated contexts.

Some researchers and skills developers do use the concept of competence in a more integrated way. Wiek, Withycombe and Redman (2011) reviewed the literature on the learning needs associated with *sustainability science and problem solving* and proposed, in summary, five sets of *key sustainability competencies*, that is, competencies that are needed in addition to the general learning needs of graduate level professionals, for work that has a specific sustainability science focus. These are:

- Systems thinking into practice
- Anticipatory competencies
- Strategic competencies
- Normative competencies
- Interpersonal competencies.

These are all needed, *in combination with each other*, to shift a situation from an undesirable state, to a desirable state reflected by a sustainability vision, significantly, around a particular intervention point (see Figure 9.1). This framing starts to reflect a situated and integrated approach to the deployment of competence, through the central role given to an 'intervention point', and the dynamic combination of diverse competencies.

In the context of leadership development that responds to complex contexts and 'wicked problems', Scharmer (2009) refers to technical, relational and

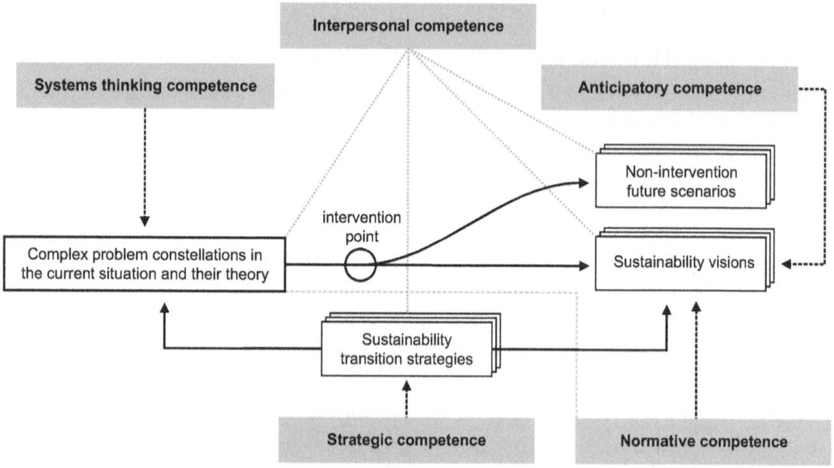

Figure 9.1 A combination of competencies are needed around an intervention point for sustainability science and problem solving

Source: Wiek et al., 2011 p. 206

Table 9.1 A matrix of intervention points (system levels) and knowledge/skill types for leadership development

Intervention points	*Types of knowledge*		
	Technical knowledge (technical skills)	*Relational (stakeholder coalition building)*	*Transformational (self-knowledge, identity, will)*
Whole system (multiple issues)	System wide technical skill building/training	System wide relational capacity building/training (multi-stakeholder dialogue)	System wide transformational capacity building (multi-stakeholder innovation)
Institution (single issue)	Institutional technical skill building/training	Institutional relational skill building/training (multi-stakeholder dialogue)	Institutional transformational capacity building (multi-stakeholder innovation)
Individual	Individual technical skill building/ training	Individual relational capacity building/ training (multi-stakeholder dialogue)	Individual transformational capacity building (multi-stakeholder innovation)

Source: Scharmer (2009)

transformational knowledge (see Table 9.1) which resides in current or future leaders (individuals) but also in institutions and 'whole system' contexts. Scharmer warned that much education provision is focused on the lower left hand corner of Table 9.1, that is, the technical training of individuals, when the resolution of wicked problems, or the taking up of innovative or transformative opportunities, requires all the forms of knowledge, at all levels depicted in the table, including transformative and whole system (top right square).

Thus a focus on the technical training of individuals would be inadequate if the intent is to develop system wide capacity for something challenging and innovative (such as a green economy transition). A leadership development specialist, Scharmer linked the knowledge required by leaders to curriculum development, but not in a narrow one-to-one relationship.

The GELA-SA adapted Scharmer and Wiek's framings to develop a frame of 'technical, relational and transformational competencies' to probe learning needs with the sustainability and green economy practitioners and education providers we consulted. We grouped normative and anticipatory competencies as 'transformational', and replaced the term 'interpersonal' competencies with 'relational competencies' to denote a broader scope (discussed further below).

In the next section I share some of the findings from three case studies undertaken in the GELA-SA for illustrative purposes.

Findings regarding the learning needs of sustainability practitioners

Functional teams in sustainable transport in urban contexts

Like other metropoles in South Africa, the City of Johannesburg had initiated efforts to address climate change and other sustainability challenges, and to reap the benefits of moving to a greener economy. In the process the municipality worked with ICLEI (Local Governments for Sustainability, http://africa.iclei.org). A project to introduce the benefits of sustainable public transport formed a GELA-SA case study to discuss what learning sustainability practitioners working in this context needed. The project, dubbed the *City of Joburg Eco-Mobility Sustainable Transport Project*, involved closing off the central business district of Johannesburg, which is the commercial and financial hub of the country, to motorised transport for a full month. It was an awareness raising exercise that required the involvement of multiple role players, and the ability to 'make the case' to a range of senior decision makers. The staff had to work effectively across institutions (local government, business community, ICLEI), across departments in one institution (traffic, environmental management, mayor's office, communications), and across levels in those departments.

Our key informant, the 'green economy champion', was the project lead within ICLEI. She explained that such projects (of which she had initiated and implemented many over two decades of working on sustainability in local government) required a team of sustainability practitioners with graduate level, multi-disciplinary areas of expertise: in engineering, economics, town planning, marketing and communications, among others. She noted that these skills were 'scarce', as not many science and social science graduates opted to work in a 'for good' context in local government.

The project lead herself had to demonstrate strong visionary and motivational leadership to inspire the team of young professionals, hold them together and provide mentoring. She noted that while the post-graduates who joined her were well qualified, they needed to learn 'on the job' how to function in a multi-disciplinary team with transdisciplinary abilities, to draw on and share their individual backgrounds and technical knowledge, but also to understand and engage the departure points of team members from different disciplines, as well as the local government officials and politicians whom they needed to engage for the successful execution of the project.

These are what sustainability practitioners often refer to as 'soft skills', to distinguish them from the technical knowledge on which a degree in systems modelling or engineering focus. Wiek at al. use the term *interpersonal* competencies, and PAGE refers to *participatory* competence, but we chose Scharmer's term *relational*. The required competence described above goes beyond being able to be personable and persuasive on a one-on-one basis, as might be inferred from the term 'interpersonal'. Rather, it involves recognising, inter alia, the value of multiple and diverse role players; being able to understand and explain the importance of the work in a bigger, multi-institutional context, and then to practically manoeuvre in that context to 'make the case' and 'take others along' towards an eventually shared outcome.

Recognising the importance of good interpersonal, inter-group and inter-institutional relationships is paramount, along with a general disposition, sensitivity and skills to establish these relationships. For post-graduates fresh from university, brimming with enthusiasm in their new knowledge and their desire to bring about urgent change, learning to 'slow down' and 'walk with others' can be vital. Our data included reports from and about young professionals working more or less alone and unable to 'make the case' to officials in local government to sustain the innovations which they had introduced through sheer hard, technical work, and who then experienced physical and psychological symptoms when they saw these efforts perishing without institutional support. But this points not only to individual experience and disposition, but to the need for enabling conditions in which the competencies of individuals can effectively contribute to the achievement of sustainability goals; South Africa's municipalities have been notoriously difficult spaces in which to do transformative work.

Renewable Energy Independent Power Producers Public-private Procurement Partnership (REIPPPPP)

The REIPPPPP is regarded as one of South Africa's most successful green economy initiatives to date. While the country's progress in introducing renewable energy had a slow, protracted start, from 2011 onwards progress accelerated and by 2014, US$14 billion had been committed to generate clean energy through the REIPPPPP. By 2015, when the GELA-SA was undertaken, the project involved more than 100 stakeholders and was regarded as the most successful public-private partnership in Africa in the last 20 years, and one of the top 10 public-private renewable energy projects globally. While external factors (like the international market for renewable energy technology) played a role in this success, the competencies of particular individuals and teams were also significant. This is evident in a report by Eberhard, Kolker and Leighland (2014), on which this case study drew.

The project management team were responsible for designing the bid process, drawing up the bid specifications, doing the bid communications and bid management, as well as overseeing the outcomes into procurement to contract completion. Several bid rounds were undertaken, each building on lessons learnt from the previous round. Eberhard et al. (ibid.) argued that the project management team displayed the following competencies:

- Coalition building – bringing together stakeholders from government and business, and establishing credibility with and winning the trust of both groups; this was achieved, for example, through professionally run bid conferences and fair bidding practice, in a context where investors were nervous about the government's commitment to the process, given, for example, the preceding long lag periods

- Championing – the team "*relentlessly*" made the case to the government and other stakeholders that this was an important initiative, and eventually won the necessary support; this reflects an ability to provide and act on visionary leadership and the qualities described as 'stamina', 'staying power' and 'tenacity'
- Problem solving and reflexivity – learning from earlier bid rounds and making important improvements, which, for example, brought down the costs; this included anticipatory skills, such as being able to anticipate counterproductive moves by bidders
- Project design and implementation in cross-disciplinary teams that had technical knowledge in, for example, Energy and Finance
- Sourcing and effectively using others' expertise: making effective use of local and international experts, which had to be carefully managed to make sure that the project received exactly what it required.

In addition to these relational and transformational competencies, including the ability to obtain and effectively use technical support from a range of experts, the REIPPPPP team also needed their own technical competencies, such as forecasting, risk assessment, contracting, financing and budgeting and setting up functioning units.

According to Eberhard et al. (2014) a critical success factor of the REIPPPPP was the setting up of its project management unit *outside* the government departments, where it did not have to follow the Public Finance Management Act and Treasury's extensive supply chain procedures. This allowed them to move with the necessary speed, but also opened the unit up to high levels of scrutiny and, potentially, allegations of reckless expenditure. The staff therefore had to act in a highly principled manner, deriving their own shared rules. The need for the ability to interpret guidelines at the level of principle and not at the level of procedure, was a finding across case studies. What it meant was that teams needed the competence to produce and follow their own, scrutinisable ethical leadership, instead of the competence to follow instructions. It is an example of normative and transformative competence.

Eberhard et al. (2014) also flagged areas in which the REIPPPPP lacked the required competencies. The government required the power providers bidding to set up South Africa's new renewable energy plants, to include a social development component in the bid and to ensure that local residents derive employment and other benefits from the solar and wind farms being erected in their area. The impact on social development also had to be monitored. Eberhard et al. (2014) concluded that few of the bidders (technical energy experts) had the necessary social development expertise to produce sound social development plans as part of their bids; the REIPPPPP team also lacked the capacity to strengthen that component of bids; and there might also not be the necessary expertise to appropriately monitor and evaluate the social impacts in a reflexive manner that would not only help to iron out implementation difficulties as they unfold, but also to provide system wide learning that could guide further renewable energy initiatives in an area where learning is clearly needed.

The importance of competence in *reflexive monitoring and evaluation* also emerged in the case study on water management, and is discussed next.

Strategic adaptive management for natural resource management: The Working for Water Programme

Working for Water is an ecological restoration programme with a jobs and poverty relief focus. As part of the Public Works Programme discussed in Chapter 6, it has been sustained over 20 years, during which time it also spawned several other natural resource management projects (such as wetland restoration and coastal care). Its budget exceeds any other natural resource management project in South Africa. This GELA-SA case study focused on the high level skills that enable such prolific initiatives.

It was significant that the programme started with effective collaboration between civil society, scientists and government, confirming the need identified in the previous two studies for cross-institutional transformational and relational competencies: where multiple role players are able to envisage something new and to work together to make it happen. Also important was the 1998 National Water Act which enshrined both human rights and environmental rights, while amendments to the longstanding Conservation of Agricultural Resources Act (CARA Act 43, 1983) were also relevant, contributing to programme traction and sustainability.

As at any one time Working for Water has more than 300 projects active across South Africa, creating employment for approximately 20 000 low skilled people per annum, and receiving commensurate budgets, the programme managers clearly have the ability to 'make the case' to their funding sources that it should be supported. '*Making the case*' for sustainability and green economy initiatives, which often translates into making a business case *as well as* a social justice case, was evident as a key learning need across the GELA-SA.

As in the previous two case examples (of a sustainable transport initiative and renewable energy procurement), it is not only or even primarily individual expertise that is necessary, but competence distributed across functioning units at various scales and levels. For example, bigger initiatives within the programme experience challenges associated with centralised decision-making and limited accountability for specific goals at lower levels; smaller initiatives with a greater degree of decentralised decision-making and shared accountability were reported to work better. This suggests that organisational and project design competencies (and not just project management skills) were important. Strategic planning was reported as a challenge; planners struggle to integrate across diverse goals (ecological, social, economic, political – giving rise to the need for *integrated sustainable development competencies*) and *stakeholder engagement and collaborative planning* again stood out as key learning needs.

Project management, while a pervasive implementation challenge, entailed more than the basics. It included *reflexive and strategic adaptive management* so that managers at various levels (from supervisors leading teams of workers, to head office departmental heads and managers in-between) can make strategic decisions based on changing or unexpected conditions and institutional emergence. Whereas the hierarchical structures and compliance procedures that characterised the government context of the programme called for the ability to follow instructions, an inability to reflexively feed into the structures and procedures, and adjust them where necessary, was a significant challenge requiring the need, again, for competence in *reflexive monitoring, evaluation, reporting and learning*, at various levels, and competence in strategic operational and organisational adaptive management.

Discussion on learning needs assessments

The GELA-SA methodology gave rise to a wealth of research findings, which can be grouped into 'competency clusters' (cf. PAGE, 2016, pp. 27–32) viz:

- Making the case for a green economy initiative or intervention
- Integrated sustainable development planning, policy development and governance
- Strategic adaptive management and expansive or transformative learning
- Working effectively across units within and across organisations and sectors
- Working with multiple knowledge fields (practically, politically and conceptually)
- Human capacity development and organisational development
- Principle-based leadership.

Each of these clusters entails a number of *multifaceted* competencies on a spectrum of more or less technical; more or less relational; and more or less transformational. The three competency categories strongly overlapped, for example, even technical competencies (such as drawing up a green procurement strategy) required vision and working with others, that is, transformational and relational competencies were never far from the technical skills. The Scharmer derived competency framework we used for the GELA-SA was nonetheless useful as a prompt, encouraging interviewees and survey respondents to look beyond the perhaps obvious and reflect more deeply on the kinds of skills needed in their work. The notions of transformational and relational competencies helped bring greater nuance and depth to the commonly used parlance of 'soft skills', which were universally acknowledged as vital but generally poorly differentiated. This demonstrates the value of analytically separating out the dimensions of a phenomenon (competence) even when it is in reality (ontologically) inseparable, provided one does not assume that it would be easy to achieve such a separation in practice.

Across the case studies, survey and interview data, there was a considerable degree of congruence. This was particularly the case with generic skills, which were technical (such as financial management or contracting); relational (such as stakeholder engagement) and transformational (such as strategic planning). Given this congruence, it was relatively easy to summarise the findings on a spectrum of common green economy learning needs. However, there were clearly also some context-specific learning needs. For this reason, we developed a decision-making template that teams can use to check and document their context-specific competency needs, followed by a self-assessment to determine their strengths and their gaps in relation to these competencies across the team or project.

The findings provide support for the contention by Lotz-Sisitka (2018) that the competencies necessary for sustainability policy actions occur in a distributed format among collectives of professionals and other practitioners. Few if any individuals had all the knowledge and skills required to undertake the complex work required to advance the green economy in a particular context. The green economy actions reported on were driven by groups whose members worked well together, whether as management units or as in-house teams, as in-house teams working with other internal teams or with external partners and/or consultants, resource persons and advisors. To call individual sustainability practitioners 'champions' or 'pioneers' recognises the importance of agency, but also obscures the collective nature of that agency, which it is clearly important to understand at greater depth if we are to plan interventions that adequately build agency. Vandenburghe (2014, cited by Lotz-Sisitka, 2018, p. 67) provided an in-depth analysis of spokespersons in social movements in which, he claimed, "the network is collapsed into the person who embodies it ... an individual comes to personify a collective or unit". This sociological explanation of agency complements the psychological ascribing of competence to individuals. With a realist lens, Lotz-Sisitka (2018) also pointed to the value of Vandenberghe's analysis (2014) of the conditions of the possibility of the group and the champion emerging and how this is mediated – the historical and socio-material factors that shape possibilities of an individual exercising agency on behalf of the group. Examples include cultural commitments to mediating leadership.

Conclusions

Green skills researchers wanting to explore learning needs (e.g. in training or learning needs assessments or skills audits) require a conceptually strong framing for the kinds of learning outcomes associated with green work and sustainability transitions in complex situations. If framings of concepts like competence (as well as skill) are narrowly reductionist, they will fail to match the complexity of sustainability work. How we frame learning needs (as a unit of analysis) is important, as it shapes the questions we ask and the extent of what this enables us to see. Simply identifying lists of competencies or skills needed by individuals is unlikely to fully reflect the learning required for complex sustainability practices in the workplace. Lotz-Sisitka (2018) argues that this use of 'competence' gives rise to a flat ontology, as the depth considerations of how the competencies came into being are missing.

The GELA-SA showed that the circumstances surrounding particular policy practices play a vital role in the success of the practitioners' endeavours. The global market forces benefitting the renewable energy programme, or new legislation, are examples. Policy practitioners' competencies are enabled by conducive conditions, even as they act to create such enabling conditions.

A realist perspective (Lotz-Sisitka, 2018) encourages researchers to look not only at individuals and their skills needs but at their interactions and the multiple interacting layers of their context, such as history, institutions, cultures and other structural factors which, as mechanisms of emergence, conspire with the agents to produce certain outcomes. Neglecting this multiplicity of factors from which green economy transitions may emerge can lead government, funders and providers to assume that short courses and other technical interventions focused on the cognitive abilities of individuals would be all that is required, thus falling into the trap mapped out in Scharmer's matrix (Table 9.1). Instead, it would be more effective to have joined-up initiatives that build the capacity of practitioners at the same time as they create more enabling conditions (cf. the notion of social ecosystems in Chapter 8).

Lotz-Sisitka (2018) argued, along with Wiek et al. (2011; cf. Figure 9.1), that competence is called forth in relation to an intervention point, or shared object comprising a sustainability goal or associated tasks. Competencies then become evident in shared work practices. As several other studies noted (cf. Chapter 5), green jobs are done by occupational networks or 'functional units' who work relationally on challenging and often new undertakings.

The nature of this shared green work is captured by the philosophical framing of ontological, epistemological and axiological (moral and teleological) dimensions. Critical realism helps us think of the complex undertakings towards a green economy as transformations in the ethical-ontological sphere: something is wrong (climate change due to polluting energy generation conventions) and based on strong shared values (desiring healthy ecosystems and thriving societies), governments and citizens wish to effect changes from what is to what should and could be; such changes may take the form of absenting absences (the absence of clean energy generation), or absenting the constraints that keep certain social-ecological ills (absences) in place. The distributed agency that is needed across individuals, organisations and systems to tackle these challenges can be described as multifaceted and overlapping technical-relational-transformational competencies, or as "practical [including relational] agency, which entails moral and epistemological knowledge that informs it" (Norrie, 2010, p. 28).

This framing for learning needs has implications for how we study competencies and enquire into learning needs. It also has implications for the design of curricula and programmes offered at university and elsewhere for graduates, undergraduates and established and early career professionals. First, in-depth and high levels of basic skills are needed, and most likely will be found across individuals. Each individual therefore needs a significant period of study and practice to build up the depth and breadth of knowledge required, so time is needed. However, they also need the ability to recognise the value of other knowledge forms, disciplines, forms of enquiry and practice. This appreciation for the wider body of knowledge and practice beyond one's own field is an important part of a basic education, and of specialisation.

Secondly, most courses, and especially short courses, are at best designed to raise awareness or convey technical knowledge. The respondents in this study noted that while such training opportunities were useful, this was the case only if short courses were complemented with:

- In-depth basic degrees and specialisations (powerful knowledge in Wheelahan's (2007) terms)
- On-the-job training and learning
- Mentoring of various kinds including peer learning
- Learning networks in which professionals support each other as they encounter new challenges.

Chapter 10 takes further the discussion on curriculum design and pedagogy to optimise the variety of ways in which sustainability students and practitioners learn. Such design features include course activated learning networks with change projects in which learners themselves have an opportunity not only to engage with given knowledge, but to co-construct new knowledge (expansive social learning) as they engage, through reflexive action at the 'intervention point', with the deeper structural layers shaping their collective agency.

Acknowledgements

The research undertaken by Eureta Rosenberg (Rhodes University), Garry Rosenberg, Nina Rivers, Presha Ramsarup (Wits University) and Heila Lotz-Sisitka (National Research Foundation/Department of Science and Technology SARChI Chair in Global Change and Social Learning Systems) in producing the *Green Economy Learning Assessment: South Africa* is acknowledged, as well as the Leadership in PAGE of Najma Mohamed (ILO) and Amrei Horstbrink and Angus Mackay (UNITAR). The study was funded by UNITAR and the South African government participation in PAGE was led by Jenitha Badul (Department of Environmental Affairs).

References

Eberhard, A., Kolker, J., & Leighland, J. (2014). *South Africa's Renewable Energy IPP Programme: Success factors and lessons*. Washington, DC: World Bank Group.

Gamble, J. (2018). Chapter 2: From labour market to labour process. Finding a basis for curriculum in TVET. In S. Allais & Y. Shalem (Eds.). *Knowledge, curriculum, and preparation for work*. Rotterdam and Boston: Brill/Sense.

Lotz-Sisitka, H. (2018). Think piece: Pioneers as relational subjects? Probing relationality as phenomenon shaping collective learning and change agency formation. *Southern African Journal of Environmental Education, 34*, 61–73. doi: 4314/sajee.v.3:1.5

Norrie, A. (2010). *Dialectic and difference. Dialectical critical realism and the grounds of justice*. London and New York: Routledge.

Partnership for Action on Green Economy (PAGE), Department of Environmental Affairs (DEA), United Nations Institute for Training and Research (UNITAR), & Rhodes University. (2016). *Green Economy Learning Assessment South Africa: Critical competencies for driving a green transition*. Pretoria: Department of Environmental Affairs/PAGE.

Republic of South Africa (RSA). Conservation of Agricultural Resources Act (CARA). (1983). Act 43 of 1983. South Africa, Pretoria: Government of South Africa.

Republic of South Africa (RSA). National Water Act. (NWA) (1998). Act 36 of 1998. Pretoria: Republic of South Africa.

Rosenberg, E., Lotz-Sisitka, H.B., & Ramsarup, P. (2018). The green economy learning assessment South Africa: Lessons for higher education, skills and work-based learning. *Higher Education, Skills and Work-Based Learning*, 8(3), 243–258. doi: 10.1108/HESWBL-03-2018-0041

Scharmer, O. (2009). *Ten propositions on transforming the current leadership development paradigm*. Presented at the World Bank Round Table on Leadership for Development Impact, Washington, DC.

Vandenberghe, F. (2014). *What's critical about critical realism? Essays in reconstructive social theory*. London: Routledge.

Wheelahan, L. (2007). How competency-based training locks the working class out of powerful knowledge: A modified Bernsteinian analysis. *British Journal of Sociology of Education*, 28(5), 637–651. doi: 10.1080/01425690701505540

Wiek, A., Withycombe, L., & Redman, C.L. (2011). Key competencies in sustainability: A reference framework for academic program development. *Sustainability Science*, 6(2), 203–218. doi:10.1007/s11625-011-0132-6

10 Green skills supply

Research from providers' vantage point(s)

Heila Lotz-Sisitka

Introduction

Environment and sustainability issues are cross-cutting, and their management and engagement requires a cross-cutting, cross-sectoral response. Primary and secondary industries and their associated sectors all have environmental management responsibilities under the National Environmental Act (1998), as can also be seen in the introductory chapters of this book (cf. Chapters 1, 2, 3, 4 and 5). Environmental management and emerging green economy responsibilities at the local level (for example, waste management and water resources management) are integrally linked to municipal service delivery and thus to local government. Management of biodiversity and coastal zones and resources are the responsibility of conservation organisations and provincial government authorities, while wider, more strategic issues such as climate change adaptation and sustainable development of the country are national responsibilities. Environmental health is a community issue and environmental concerns are also integral to poverty alleviation especially in rural areas, where communities rely heavily on direct access to, and use of natural resources for their health and well-being. Environmental skills are therefore cross-cutting, and are needed in:

- Primary and secondary industries and all sectors of the economy
- Local government management practices within integrated development planning systems
- Provincial governance across government departments and sectors
- National governance across government departments
- Community development contexts in a range of settings (urban, rural, peri-urban)
- The formal education sector (which builds the foundations of nationally relevant, cross-sectoral and foundational skills, knowledge and values in which environment and sustainability are cross-cutting concerns that are relevant to a range of subjects as has already been foregrounded by the three iterations of curriculum policy in South Africa).

The cross-sectoral nature of environmental skills development and the complexity of planning for cross-sectoral skills of this nature imply inter-disciplinary foundations for environmental skills development. This has implications for the way in which providers respond to the emerging landscape of green skills development. It is interesting to note that, in recent years, the majority of the Sector Skills Plans in South Africa indicate that Sector Education and Training Authorities (SETAs) have expanded their interest in green skills supply, but in most cases the commitments to green skills remains at the level of research. This may, in part, be due to an inadequate providers' capacity for green skills supply, which is also one of the findings in research reporting on skills development in the Expanded Public Works Programme's training initiatives (cf. Chapter 8). The Environmental Sector Skills Plan (Department of Environmental Affairs [DEA], 2010) in South Africa pointed to a problem of over-emphasis on High Skills and specialisation in the environmental sector, which is also reflected in supply side data showing this bias (Ramsarup, 2017), and Ramsarup (2016, 2017) identified the problem of a 'missing middle' in the supply of green skills in the environmental sector in her review of qualifications on offer for green skills development in South Africa. The Environmental Sector Skills Plan for South Africa (DEA, 2010) and other human capacity reviews have all pointed to the importance of building foundational knowledge in the school, college and university system, and among the public where environment and sustainability literacy is crucial for the advancement of a more sustainable, just society.

Given the above, there is a huge need to focus on the supply side of skills development, especially to balance the relationships between supply and demand in the sector, and to improve the quality and scope of the supply of green skills to the economy, but also more broadly to society, so as to ensure 'green economies and societies', not just green economies (cf. Chapter 1). As shown in the chapters across this book, more refined analysis of environmental sector skills demand raises new and interesting perspectives on skills supply, and the issue of curriculum responsiveness arises, as does the issue of wider social learning. Thus supply side questions that are relevant to the advancement of green skills in South Africa include a focus on inter-sectoral skills demands and their specific nuances, as well as the inter-disciplinary foundations required for inter-sectoral and systemic responses to address environmental and sustainability concerns in the country, and growing a green economy within a just transitions framework. These concerns also raise questions on the role of transformative social learning more widely, and how such processes are to be constituted, supported and co-engaged.

In recent years, there have also been numerous discussions about inter- and transdisciplinary orientations to green skills development, given the need to deal with more complex social-ecological system concerns, as was shown in a regional study by the Southern African Regional Universities Association on how universities across the Southern African Development Community (SADC) region were responding to climate change (Urquart & Lotz-Sisitka, 2014). In the section below, I offer insights into four research project examples that can facilitate a more systemic, inter-disciplinary and transformative orientation to skills development research, conceptualised from a providers' vantage point. The case studies focus on skills development at different levels and within different accreditation frameworks, the QCTO (Quality Council for Trade and Occupations), the CHE (Council of Higher Education) and beyond these boundaries, to show the diverse demands involved in contexts of accreditation, supply and provision.

Research from a provider's point of view

Most research into skills supply and provisioning in South Africa is conducted within the wider context of a fundamentally skewed and unequal system of education and training, and a history of poor quality education and training (Motala & Vally, 2014). This, combined with the 'newness' of the green skills field in South Africa (cf. Chapters 2 and 14), creates a very challenging context within which green skills supply and provisioning research takes place. As mentioned in Chapter 2, prior to 1994, there were initially very few green skills programmes, while historical models of education and training, given their histories of colonial construction and marginalisation via poor quality, are 'mismatched' with the kind of transformative learning that is required in the environmental sector and for green skills development. This has resulted in a need for green skills supply that focuses both on quality of provisioning, as well as the scope and and diversity of education and training contexts.

In this section I share some examples of green skills supply research which illuminate how such research processes can be constructed, what methodologies can be used and what types of associated findings emerged from these studies. These studies are briefly introduced and readers are referred to the referenced material in each case for more insight into the studies. There are a range of other green skills supply studies that can shed further light on green skills supply issues, and that can inform various dimensions, such as course and curriculum design, learning materials development and use, transformative pedagogical research, assessment and other concerns such as the inter- and transdisciplinary dimensions of green skills supply. There are also studies that shed light on the particular challenges that emerge in the different sub-sectors of the national system of skills supply (for example, technical and vocational education and training, early childhood education, specialist higher education training, etc.) (for some examples, see Price and Lotz-Sisitka, 2017 and the *Southern African Journal of Environmental Education*, www.ajol.info/index.php/sajee).

These studies are important for the perspectives provided on *how to* go about developing new training programmes, or on how to improve the quality of education and the types of situated, transformative learning that are needed for green skills development in South Africa conceptualised within a framework of just transitions or transformations to sustainability. Of interest to this book is the fact that most of these studies move beyond the traditional 'training needs analysis' (cf. also Chapter 9) and instead seek deeper understandings of the complex dynamics of skills supply and provisioning.

Case study 1: Understanding the workplace context to inform curriculum design

One of the challenges of designing new qualifications or courses is to have a good understanding of the context in which the training programme is to be implemented. One of the most challenging areas for green skills supply in South Africa is the local government context, where environmental and sustainability issues are a core concern for efficient service delivery, and in 2010 the Environmental Sector Skills Plan for South Africa indicated that there were over 30 000 people in South Africa working in environmental occupations in this sector, yet there was almost no training on offer for them. This led to the Local Government Sector Education and Training Authority identifying the need for Environmental Practices Training at Levels 2–4 (elementary occupation level). While the Level 2–4 training demands for environmental competence were relatively easy to identify, it was less easy to understand who would conduct the training in these levels, and therefore also what competences were required for Level 5 environmental educators. In 2006, a process was started to inform the development of qualifications for Environmental Education, Training and Development Practitioners. Wigley (2006) undertook research to understand the workplace epistemologies (ways of knowing) and institutional cultures of Level 5 trainers in local government contexts and in agricultural training contexts. His study developed an understanding of workplace epistemologies related to the environment and education, showing the inter-disciplinary nature of the field, as well as the structural factors that enable and constrain agency of environmental educators and the role of reflexivity in practice and in education.

The study revealed diverse and seemingly ambiguous understandings of both environmental/sustainability and education processes in the two contexts. This ambiguity seems to relate to environmental education practitioners drawing on different forms of knowledge, including differentiated or theoretical knowledge and 'common-sense' ways of knowing, in their education practice. The understandings related to theoretical knowledge were, in both contexts, dominated by scientific or technical understandings where the environment is understood in terms of the natural sciences and education is seen in instrumentalist terms as the transfer of mainly technical environmental knowledge to learners in order to effect behaviour change. This showed that the cultural histories that shaped education in the past produced continuities into the present, and that transformative orientations for green skills development therefore needed to challenge these historical cultures in both the environmental field and the educational field. As a result of this finding, the study went further to examine epistemological, socio-cultural and structural features of contexts in order to better understand how these historical structures shaped the possibilities for transformative education. Interestingly, the study found that it was not only the historical cultures of the workplaces that shaped environmental educators' orientations but also that the intentionalities of the educators themselves were a significant shaping factor determining how the training was conceptualised and offered.

The study argued for strengthening the reflexivity of the environmental educators, and concluded that a course for environmental educators at Level 5 should include a strongly reflexive approach to course design and delivery. This study therefore developed an understanding of curriculum as a contextualised social process that involves structural aspects of the curriculum such as institutional cultural histories, as well as socio-cultural processes like reflexive learning on the course and in the workplace.

Case study 2: A whole university sustainability assessment using a unit-based sustainability assessment tool

A second challenge in providing green skills is to consider the way in which sustainability is integrated into whole institutions. It has been found by the United Nations Educational, Scientific and Cultural Organisation (UNESCO), Creech and Buckler (2014) that a whole institution perspective offers students a more holistic experience of sustainability and leads to stronger inter- and transdisciplinary orientations to sustainability education. There have been substantive initiatives internationally, especially at university level, to conceptualise ways of supporting universities as institutions to adopt more holistic approaches to dealing with sustainability concerns. This is because failure to do so could leave students with contradictory forms of education. Giving attention to whole institution approaches also addresses the curriculum in its fullest sense, which is inclusive of the explicit and implicit curriculum (i.e. students learn as much from the way in which the waste is managed in a classroom or institution as they do from a lesson on waste management). One of the ways of supporting providers to see the scope and quality of their green skills offerings at whole institution level is to use an institutional audit tool. There have been a number of different university-based sustainability assessment tools produced and in her research Togo (2009) discovered that these were difficult to use in most African university contexts, because they tended to lack structures for co-ordination of environmental and sustainability concerns. In response, she designed a Unit-based Sustainability Assessment Tool (USAT) (Togo & Lotz-Sisitka, 2009, 2013) testing it in one institution in depth, and then more widely in a range of other universities (Togo, 2015).

Togo (2009) used systems thinking to guide the generation of empirical data across departments and units in the university, including student activities, policy, curriculum and teaching approaches effectively viewing the system as being composed of many parts responsible for different functions, also seeing the university as a sub-system of the wider environment in which it is situated. According to Banathy (1992, p. 22) the systems-environmental model is "a lens that projects a bird's-eye view of the landscape in which the system is embedded", which enables "the researcher to describe the system within the context of its environment" (Togo, 2016). One of the shortfalls of mainstream systems thinking is its reliance on empiricism, which does not allow "transfactual theories of causality in complex and messy open social systems where a variety of causes can lead to similar outcomes" (Togo, 2016, p. 87). Systems thinking also fails to include adequate historical and cultural emergence analysis. For these reasons, Togo underlaboured systems thinking with critical realism in her green skills research, as this allows researchers to examine systems at the level of empirical experiences, as well as the actual level of flows and consequences, and the real level of objects, their structures and powers (cf. Chapter 13). Importantly, critical realism seeks explanations beyond empirical evidence, enabling researchers to find explanations of causal mechanisms that shape human experience (Danemark, Ekström, Jakobsen, & Karlsson, 2002). Bhaskar's (2010) 'laminated system' involving the sub-individual psychological level, the individual or biographical level, micro-levels of institutional organisation, the meso-level of the student in an unsustainable society, the macro-level of the whole society or region; the mega-level of whole traditions and the planetary (or cosmological) level concerned with the planet (or cosmos) as a whole (cf. Chapter 13) were all relevant to understanding sustainability responses in the university (Togo, 2016). The university sustainability initiatives at the empirical level (for example, curriculum innovations in departments or student environmental initiatives) were analysed first, and critical realism's depth ontology (i.e. its laminated system) was used to guide explanation of what was found (e.g. contradictory interpretations of sustainability from a diversity of disciplines). Historical explanation was also used to explain how these empirical experiences came into being, which helped to shed light on causal phenomena, even where the causes were not immediately obvious.

This type of analysis showed how the university was faced with integrating sustainability challenges of its environment into its curricula, departments, research, policies, operational management and student activities, or not. The main insight from this research was that universities and other educational institutions need to depart from reductionist approaches when integrating sustainability and environmental concerns, as individual disciplinary innovations on their own are inadequate both for responding to the wider scope of sustainability concerns and for developing a fuller understanding and experience of sustainability in educational settings. Without embracing a whole institution approach that embeds sustainability into all aspects of the institution's programmes and operations, the institution is likely to only achieve a 'limited adaptive response' (Sterling, 2004, pp. 50–51). Togo's work, drawing on critical realism, shows, however, that a fuller, more holistic response to sustainability within institutions is also an emergent process, and that there is need to take into consideration that different faculties, departments and units in a university are likely to have different structures, histories, priorities, cultures, resources, leadership styles, values, visions and philosophies. This includes the way they understand and embrace sustainability (or not), and that they may all influence the emerging system in unique ways. It is therefore more helpful to adopt an approach to whole institution development research that helps to clarify these processes of emergence and/or stasis, and to generate data that can facilitate ongoing dialogue and engagement around sustainability as a common concern for humanity (i.e. planetary level concern) that has its particular features at concrete levels (ibid.). Sustainability in such a context can be viewed as a 'concrete universal' (Bhaskar, 2010) that is characterised by a diversity of forms and processes of emergence.

Examining sustainability in this way can help to make it clear that educational institutions that provide education and training for sustainability and green skills are embedded in, and related to the wider environment to which they are able to respond, albeit in a diversity of ways. Togo found in her study that the causal factors shaping the emergence of sustainability and green skills education programmes are more complex and deeper than "would have been reflected by information established through systems theories only, which was limited to the empirical level of reality" (Togo, 2016). She argued that "systems theories can therefore benefit from ontological theories of emergence that take account of a 'laminated totality' as described by Bhaskar". Her initial research and propositions on the emergence of whole institution approaches to sustainability in universities (Togo, 2009) has been validated by wider applications across over 35 other universities in Africa (Togo, 2015). Importantly, the use of the USAT tool in these universities was shown to provide tools for reflexivity that shaped emergence of new sustainability courses, curricula, programmes, policies, management activities and student engagement across these institutions.

Case study 3: Vocational education curriculum reviewing using an adapted USAT tool

The approach developed by Togo (2009, 2015, 2016) elaborated on above has been adapted and used by a variety of providers for different purposes as noted above. One such approach that has the potential to support curriculum innovation is adapting the USAT tool approach specifically for curriculum review. In one such study, Van Staden (2018) developed an adapted version of the USAT tool via an iterative, engaged research approach with agricultural college lecturers over a period of a year in which they reflected on the manner in which climate change concerns were integrated into their curricula (or not). Her concern was that while climate smart agriculture was identified as being an important driver for agricultural system innovation, few colleges had given attention to this. Early in the study, Van Staden (2018) undertook contextual analysis using a survey and document analysis to establish the 'status quo' of climate smart agriculture integration into the agricultural technical and vocational education and training (TVET) curriculum in two colleges. Using consultative workshops which included both agricultural TVET lecturers and provincial policy actors, she shared insights from the initial contextual profiling, and co-designed a set of questions to use for curriculum innovation reflection. She then used these, together with insights from a literature and policy review of climate smart agricultural principles, to develop a Version 1 of an adapted USAT tool that focused on the potential integration of climate smart agricultural principles, content and competences into the agricultural TVET curriculum. TVET lecturers completed the Version 1 USAT tool. Data generated from the Version 1 review informed further consultative workshops and a process of co-engaged reflexive review on the findings, which initially showed very little attention to climate smart agriculture in the agricultural college curricula. This helped Van Staden (2018) to build a picture of the activity systems in the college that were influencing the integration (or not) of climate smart agricultural principles and practice. Using this as mirror data and drawing on expansive learning principles from Cultural Historical Activity Theory (Engeström and Sannino, 2010), she continued to engage with lecturers around the tensions and contradictions they were experiencing in integrating climate smart agriculture principles and practices into their curricula. Findings from this phase informed a re-design of the USAT tool, and the tool was further refined into a more differentiated (for lecturers in Plant Production, Irrigation, Animal Production and Agricultural Management) and responsive online format that allowed lecturers to review their own practice, and to also find useful knowledge resources that could be used in response to areas where they were struggling (http://amanziforfood.co.za/csit-resources).

Van Staden's (2018) study models the process of reflexive curriculum review often needed for green skills supply curriculum innovation processes, in which the climate smart agricultural tool (the adapted USAT) provided a double stimulation tool, meaning it was useful for people to inform reflexive decision making and achieve resultant expansive learning outcomes in the form of transformative agency of the lecturers (ibid.; Engeström & Sannino, 2010). The study worked well at the micro-level of supporting agricultural TVET college lecturers to reflexively review, and expand their practices in response to the challenges of climate change in the agricultural sector. The study also showed that a concept such as 'climate smart agriculture' can initially be very abstract, and that iterative and reflexive engagements in the formative intervention research process can assist with enabling the concept to become more concretely understood and applied in agricultural TVET settings. This shows that curriculum innovation processes are also transformative learning processes for the lecturers concerned.

Case study 4: Transgressive, transformative, transdisciplinary (T-learning) as a focus for curriculum innovation and green skills provisioning

A fourth challenge in providing green skills research lies in the type of learning on offer. Traditionally, educational institutions have tended to focus on types of learning that foreground the cognitive gain of individuals. Most curricula and assessment systems are structured with this objective in mind. In recent years, given the scope and wicked nature of the environmental and sustainability challenges being experienced in societies around the world, including South Africa, there has been a realisation that learning that focuses on individual cognitive gain only, even if it involves 'perspective shifts', is not adequate for responding to the challenges of sustainability, or for producing alternative practices and futures. Consequently, there has been a burgeoning of interest in transformative learning, with the Intergovernmental Panel on Climate Change (2014) noting that learning is one of the cornerstones of societal responses to climate change and other major social-ecological challenges (see Chapter 1). Yet not enough is known about the type of learning required, or how such learning is to be supported and integrated into existing educational systems and approaches and/or how such learning can transform educational institutions (Lotz-Sisitka et al., 2015; Macintyre et al., 2018). A recent research project led out of the Global South into T-learning (meaning transgressive, transformative and transdisciplinary types of learning) has provided some insights into the types of learning required for the building of sustainable societies and transformative green skills learning pathways (Lotz-Sisitka, 2019). Working with and in civil society, youth, academic, government and community partners – including a variety of educational institutions – in nine countries in diverse areas that are vulnerable to arising impacts at the climate-energy-food-water security and social justice nexus, the project aimed to: 1) initiate, 2) frame and 3) investigate expansively and 4) transgress the status quo via T-learning. Each of the cases worked at the process of developing generative T-learning methodologies. These generally took the form of first co-defining matters of concern, which included analysis of cultural histories, social-ecological systemic factors and histories, political ecology and economy dynamics, as well as relational dynamics. From here catalytic potential activities were identified for expansion that were of interest to the communities involved (for example, water harvesting and management, preservation of sacred natural sites, cultural-ecological tourism, food sovereignty practices and more), which was done via a series of co-engaged and contextually defined learning actions that were uncertain, iteratively emergent, reflexive and took account of the principle of 'learning what is not yet there'. This involved unlearning current cultural praxis that constrained change, and co-learning new possibilities in open systems and recognition of the unknown/excess that cannot be pre-determined or controlled.

Methodologies were also co-developed. Some took the form of change, T- or challenge LABS (forms of 'living labs/studios') meaning that people agreed to co-learn while trying to co-develop new activities together in a real life setting. Examples included productive demonstration site development, surfacing silent knowledges and exchanging knowledge and praxis across sites. It also included mobilising creativity and re-imagining the present with arts-based approaches that included developing cartographies, video, podcasts, theatre pieces, song and counter-hegemonic participatory mapping praxis. Co-developing and using ICT tools such as mobile applications, social media tools like WhatsApp and participatory courses (combining online and offline) were also tried.

Importantly, the methodologies themselves were not reduced to 'techniques' but were constituted as deeply embedded social-sustainability learning processes that emerged via reflexive and ongoing transgressive co-engagements over time. A feature of these methodologies was their capability for network building and relationality, as well as critiquing, de-colonialising and challenging the status quo, while also enabling the emergence of desired alternatives as 'transgressive moves', ultimately critiquing the very notion of 'methodology'. In all cases, researchers have recognised that this work requires deep levels of engagement and critique, recognition of diversity, ethics and empathy *and* willingness to co-engage in ongoing processes of reflexivity and transgressive change (spiralling reflexivity) with deconstruction of the status quo being a foundation for building alternative praxis and change, emphasising the processes that enable the latter to emerge. The decolonial dialectic of absence and emergence (De Sousa Santos, 2014) was therefore a key feature of the methodologies, as was collective agency formation, in pluralist, multi-actor epistemic environments, where the possibility for transgression of 'norms' or 'normalised praxis' at epistemic and ontological levels was always possible. Approaches to capture such transgressions were also developed.

Interestingly for sustainability transitions, the research saw examples of practice development and transgressions of the 'norm' in transformed praxis and relations that were actualised at niche level, as well as in networked, multi-levelled formations where these were in focus. Examples of such outcomes are: previously marginalised women in Malawi were able to assert the value of their indigenous knowledge for enhancing food sovereignty, black youth in South Africa were able to assert and articulate their desires for transgressive pedagogies of the present, and severely marginalised women farmers were able to produce food and engage in food sovereignty activism (South Africa), while scientists and farmers were able to engage with communities for climate change adaptation in Vietnam, and communities in Ethiopia were able to assert their rights to conserve and protect sacred natural sites (amongst others).

From a methodological point of view, the research developed a paradigm of critically reflexive, generative and emergent approaches to knowledge co-production, substantively 'overturning' many assumptions of top-down curriculum development which is the norm in most institutions. The case sites were identified *in consultation with* participating communities and organisations, who worked *with* T-learning researchers to: 1) co-define matters of concern, 2) identify potentially catalytic focus areas or practices for further expansive learning within collaborating communities of co-researchers, 3) develop situated T-learning methodologies for learning-led change, and 4) undertake reflexive reviews of these processes of learning with an emphasis on observing and commenting on power relations, contextually relevant debates around T-learning (e.g. pluralism, decolonial sensibilities, complexity, etc.) and sustainability outcomes emerging in practice. Learning from the experience of researching T-learning across nine countries on four continents, Lotz-Sisitka (2019) reflected that:

> In this research programme, we have expanded this understanding of transformative learning to include transformations in collective relationality, collective learning, and collective agency in uncertain conditions and in open systems, in transition movements where excess also operates as a defining feature of transgression. We have provided nuance and depth to the popular notion of 'co-learning' in the Social Ecological Sciences (SES)

> from a methodological as well as theoretical perspective. We have, in particular, given more in-depth meaning to the notion of **transgressive** in the learning sciences, especially in a sustainability context, but also in a context of related social justice and de-colonial scientific contexts. We have, therefore, expanded the meaning of **Transgressive forms of Transformative and Transdisciplinary learning in the Sustainability Science context**. … in general situated, co-engaged and critically constituted methodologies that confront colonial, exclusionary, unsustainable and master-slave type relations produce qualities of T-learning that foreground cognitive and epistemic justice, environmental justice, solidarity, acknowledgement of silent and previously hidden or marginalised knowledges, empathy, emotion, relationality, anger and other qualities of learning that are often ignored in mainstream learning sciences that tend to focus on cognition only. Our research therefore draws attention to this wider range of qualities of T-learning, which equally recognize the importance of new knowledge, but also the importance of how this relates to and emerges from or 'moves with or against' other forms of knowledge such as tacit knowledge, spiritual knowledge(s), indigenous knowledge, experiential knowledge and more. It therefore offers a more complex 'ecologies of knowledge' (De Sousa Santos, 2014) configuration for qualities of learning than are recognised in the mainstream learning sciences, and in the contemporary SES [Social-Ecological Sciences] which tend to focus more instrumentally on first, second and third loop learning as main descriptor and analytical tool for learning.

The outstanding challenge is to integrate this into normalised systems of curriculum development for green skills research, although some good examples of curriculum implications also emerged in the project (Chaves, Macintyre, Verschoor, & Wals, 2018; Pesanayi, 2019).

Conclusion

The research perspectives from a providers' vantage point outlined above show the dynamic nature of supply side research in the green skills arena. The background information and the four studies shared above also indicate the demand and the scope of possibilities for green skills supply, which range from studies of how the system is incorporating green skills from a qualifications design and coverage point of view, through to an understanding of how histories and cultures of experience and institution can shape supply and provision, pointing to new dimensions for curriculum design, and reflexive curriculum review processes that also build the capacity of lecturers to understand new concepts and knowledge emerging in the green skills arena. They also show the need for considering these issues in a whole institution context, recognising that educational institutions are sub-systems or parts of a wider whole environment. Here, there is even need to probe systemic vantage points critically in order to not be left with limited views and/or 'adaptive' responses only.

Transformative responses require transgressions and transdisciplinarity; transgressing taken for granted views on learning that are limited in their ability to drive transformations to sustainability (Lotz-Sisitka et al., 2017). They also require generative ways of re-thinking learning, curriculum, whole institution development and a 'learning of what is not yet there' (Engeström, 2016). Complementary chapters in this book show that such learning approaches and curriculum interventions need to also take account of learning pathways, the contemporary nature of environmental/green career trajectories (i.e. they are often boundaryless – cf. Chapters 6, 7 and 8), transitions into green work, system elements such as occupational descriptors (cf. Chapters 5 and 12), more imaginative demand side studies that can articulate latent green jobs for green and greening work (cf. Chapters 2, 3 and 4) and the wider contributions of all of this to more just, sustainable societies overall (cf. Chapters 1 and 14).

Overall, we can argue that in supply side research, there is need to focus on a diversity of units of analysis: the individual learner's experience, the work of the lecturer in curriculum innovation, qualification, course and curriculum, the learning pathway, the department/programme and whole institution, as well as boundary crossing T-learning across organisations and a diversity of actors in society, which offers a broad canvas for supply side research in the green skills arena. There is need to examine what is present in these settings, but also what is absent so as to chart potential transformative pathways as pointed out by Ramsarup's (2017) 'missing middle' argument and by Van Staden's (2018) study shared above. Questions associated with green skills supply cannot be separated from wider contextual factors and histories, and these need to be interpreted critically within systems approaches, offering new perspectives on 'systemic articulation' in national education and training systems (Lotz-Sisitka & Ramsarup, 2017). Such a perspective can also expand supply logic to beyond the qualification, course and institution only, to include a paradigm that embraces wider social learning and boundary crossing learning. The T-learning research reported on above has taken the concept of supply and provisioning out of the course, classroom and educational institution into wider frameworks for co-engaged social learning amongst multi-actors across levels of emergence (cf. also Chapter 11), a perspective which has much to offer green skills development in South Africa, especially if it is to be inclusively conceptualised as is the intention with the notion of 'green skills for just transitions'.

References

Banathy, B. (1992). *A systems view of education: Concepts and principles for effective practice*. Englewood Cliffs, NJ: Educational Technology Publications.

Bhaskar, R. (2010). Contexts of interdisciplinarity and climate change. In R. Bhaskar, C. Frank, K.G. Hoyer, P. Naess, & J. Parker (Eds.), *Interdisciplinarity and climate change: Transforming knowledge and practice for our global future*. London and New York: Routledge.

Chaves, M., Macintyre, T., Verschoor, G., & Wals, A.E.J. (2018). Radical ruralities in practice: Negotiating *buen vivir* in a Colombian network of sustainability. *Journal of Rural Studies*, 59, 153–162. doi: 10.1016/j.jrurstud.2017.02.007

Danemark, B., Ekström, M., Jakobsen, L., & Karlsson, J. (2002). *Explaining society: Critical realism in the social sciences*. London: Routledge.

Department of Environmental Affairs (DEA). (2010). *Environmental Sector Skills Plan for South Africa*. Preoria: DEA.

De Sousa Santos, B. (2009 [2014]). *Epistemilogicas do Sul*. Coimbra: Almendina [Translated into English as 'Epistemologies of the South – Justice Against Epistemicide']. London: Verso.

Engeström, Y. (2016). *Studies in expansive learning: Learning what is not yet there*. Cambridge: Cambridge University Press.

Engeström, Y., & Sannino, L. (2010). Studies of expansive learning: Foundations, findings and future challenges. *Educational Research Review*, 5(1), 1–24. doi: 10.1016/j.edurev.2009.12.002

Intergovernmental Panel on Climate Change (IPCC). (2014). *Climate change 2014: Impacts, adaptation, vulnerability*. Cambridge: Cambridge University Press.

Lotz-Sisitka, H. (2019). *T-learning in times of climate change*. Unpublished research report submitted to the International Science Council, Paris.

Lotz-Sisitka, H.B., Wals, A.E.J., Kronlid, D., & McGarry, D. (2015). Transformative, transgressive social learning: Rethinking higher education pedagogy in times of systemic global dysfunction. *Current Opinion in Environmental Sustainability*, *16*, 73–80.

Lotz-Sisitka, H., Mukute, M., Chikunda, C., Baloi, A., & Pesanayi, T. (2017). Transgressing the norm: Transformative agency in community-based learning for sustainability in southern African contexts. *International Review of Education*, 63(6), 897–914. doi: 10.1007/s11159-017-9689-3

Lotz-Sisitka, H.B., & Ramsarup. P. (2017). Learning pathways and articulation: Early conceptual explorations and implications for research design(s). *SAQA Bulletin: Learning pathways, the National Qualifications Framework (NQF), and lifelong learning in South Africa*, *17*(1), 31–48.

Macintyre, T., Lotz-Sisitka, H., Wals, A., Vogel, C., & Tassone, V. (2018). Towards transformative social learning on the path to 1.5 degrees. *Current Opinion in Environmental Sustainability*, *31*, 80–87. doi:10.1016/j.cosust.2017.12.003

Motala, E., & Vally, S. (2014). *Education, economy and society*. Pretoria: University of South Africa.

Pesanayi, V.T. (2019). *Boundary-crossing learning in agricultural learning systems: Formative interventions for water and seed provision in southern Africa*. Unpublished PhD thesis, Rhodes University, Grahamstown, South Africa.

Price, L., & Lotz-Sisitka, H. (2017). *Critical realism, environmental learning and social-ecological change*. London: Routledge.

Ramsarup, P. (2017). *A critical realist dialectical understanding of learning pathways associated with two scarce skill environmental occupations within a transitioning systems frame*. Unpublished PhD thesis, Rhodes University, Grahamstown, South Africa.

Republic of South Africa. (1998). National Environmental Management Act. Pretoria: Government Printers.

Sterling, S. (2004). Higher education, sustainability and the role of systemic learning. In P.B. Corcoran & A.E.J. Wals (Eds.), *Higher education and the challenge of sustainability: problematics, promise, and practice*. Dordrecht and Boston: Kluwer Academic Publishers.

Togo, M. (2009). *A systems approach to mainstreaming environment and sustainability in universities: The case of Rhodes University, South Africa.* Unpublished PhD thesis, Rhodes University, Grahamstown, South Africa.

Togo, M. (2015). Development, use and significance of the Unit-based Sustainability Assessment Tool for universities in Africa and Asia. In H. Lotz-Sisitka, A. Hlengwa, M. Ward, A. Salami, A. Ogbuigwe, M. Pradhan, M. Neeser, & S. Lauriks (Eds.), *Mainstreaming environment and sustainability in African universities: Stories of change*. Grahamstown: Rhodes University, ELRC.

Togo, M. (2016). Chapter 5: Underlabouring systems thinking with critical realism in understanding Rhodes University's response to the sustainability imperative. In L. Price & H. Lotz-Sisitka, *Critical realism, environmental learning and social-ecological change*. London: Routledge.

Togo, M., & Lotz-Sisitka, H. (2009). *Unit-based sustainability assessment tool. A resource book to complement the UNEP Mainstreaming Environment and Sustainability in African Universities Partnership*. Howick: Share Net.

Togo, M., & Lotz-Sisitka, H. (2013). The Unit-based Sustainability Assessment Tool and its use in the UNEP Mainstreaming Environment and Sustainability in African Universities partnership. In S. Caeiro, W.L. Filho, C.J.C. Jabbour, & U.M. Azeiteiro (eds), Sustainability assessment tools in higher education institutions: Mapping trends and good practices around the world, pp. 259–288. Dordrecht: Springer.

United Nations Educational, Scientific and Cultural Organisation (UNESCO), Creech, H., & Buckler, C. (2014). *Shaping the future we want: UN decade of education for sustainable development (2005–2014): Final report.* Retrieved from http://unesdoc.unesco.org/images/0023/002301/230171e.pdf

Urquart, P., & Lotz-Sisitka, H. (2014). *Climate change counts. Strengthening university contributions to climate compatible development in southern Africa*. Johannesburg: SARUA.

Van Staden, W. (2018). *A review of climate-smart system innovations in two agricultural colleges in the north west province of South Africa.* Unpublished PhD thesis, Rhodes University, Grahamstown, South Africa.

Van Staden, W., O'Donoghue, R., & Lotz-Sisitka, H.B. (2018). *Climate smart curriculum innovation tool.* Retrieved from http://amanziforfood.co.za/csit-resources/

Wigley, J.J. (2006). *Understanding workplace-based learning contexts to inform curriculum development: The case of a Level 5 Environmental Education, Training and Development Practices Qualification*. Unpublished Master's thesis, Rhodes University, Grahamstown, South Africa.

11 Formative interventionist research generating iterative mediation processes in a vocational education and training learning network

Heila Lotz-Sisitka with Tichaona Pesanayi[1]

Introduction

Food insecurity is an issue in South Africa that undermines the global effort to end hunger by 2030. Recently, reports on the reduction in food availability have been noted in regions affected by El Niño phenomena including South Africa (WFP, 2016) thereby increasing the prevalence of household food insecurity which remains a major challenge in the country (Vink & Van Rooyen, 2009; Backeberg & Sanewe, 2010; Labadarios et al., 2011) and in dryland regions more broadly. While some communities in southern Africa practise subsistence agriculture, insufficient inputs, unpredictable rainfall patterns and inefficient farming practices result in extremely low yields. Therefore, it is a critical requirement to increase food availability, eliminate hunger and improve subsistence farmers' livelihoods through promotion of agricultural practices that strengthen resilience to climate change and other related stressors, especially amongst women groups, who are the mainstay of African food production systems at smallholder farmer and household food production levels.

There is a substantial body of training information developed by national research based organisations such as the Water Research Commission (WRC) in the public domain responding to the multifaceted crop production challenges faced by small-scale growers. Similarly, numerous well researched and developed knowledge resources for supporting small-scale farmers and household food producers are available. However, a major problem exists around the uptake and use of this knowledge in practice, especially in the context of the agricultural learning system, including but not limited to agricultural training institutions (Denison et al., 2011). In the Amanzi [Water] for Food formative intervention research case reported on here, we deal with this problem in the vocational agricultural learning system. However, the issue of knowledge flow is a broader issue characterising the green skills learning environment. This is because knowledge being produced for addressing environment and sustainable development issues is often 'new'; it also challenges taken for granted 'norms' of what counts as relevant knowledge in vocational and other education and training settings.

Conceptual and theoretical framework

The knowledge flow problem raises the challenge that knowledge resources cannot be simply treated as 'objects' to be inserted into a curriculum or learning system. It is widely appreciated that information (often held in knowledge resources) on its own does not lead to capability development. Education, training, knowledge dissemination and communication using different knowledge dissemination and mediation processes (Blackmore et al., 2011; Shaxson et al., 2012) are required for translating information into action. The global challenge of achieving traction from research outputs and knowledge resources is related to what is now recognised as inadequate Research-Develop-Disseminate-Adopt (RDDA) assumptions of how knowledge is/ought to be mediated in societies (Robottom, 1991; Lotz, 1995). The RDDA model has long since been critiqued for overlooking the complexity of societal relations and the structural, cultural and historical conditions that shape and influence knowledge production and dissemination (ibid.), including the mediation processes needed for knowledge engagement and uptake in social settings (Figure 11.1 illustrates the RDDA approach).

Contemporary theories of learning and change indicate that for knowledge or information to become meaningful, it needs to be related to the situation and experience of the user. Similarly, it needs to be mediated in the context of the user, providing new knowledge or information that can expand existing knowledge and/or practice. These are increasingly known as knowledge co-production processes, outlined through the use of social learning theory (Blackmore et al., 2011; Shaxson et al., 2012; cf. Chapters 8 and 10) and more systemic ways of thinking about knowledge flow (see Figure 11.2).

Social learning theory and knowledge co-production processes are being developed rapidly in adaptive natural resource management contexts and

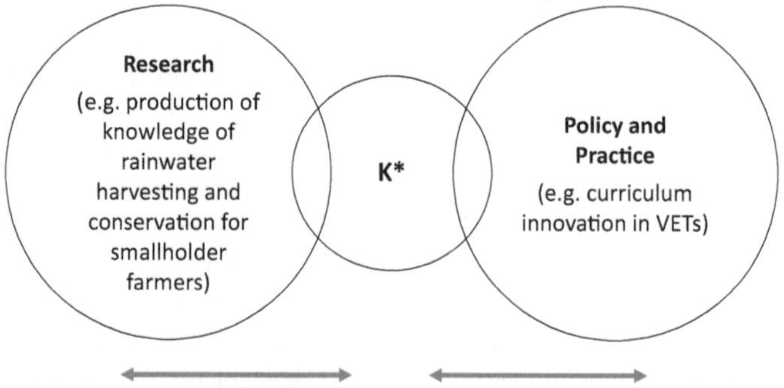

Figure 11.1 Earlier technology transfer views of knowledge dissemination and use

Note: The * indicates the variety of knowledge dissemination processes that may take place in any given context at the nexus of policy, practice and science

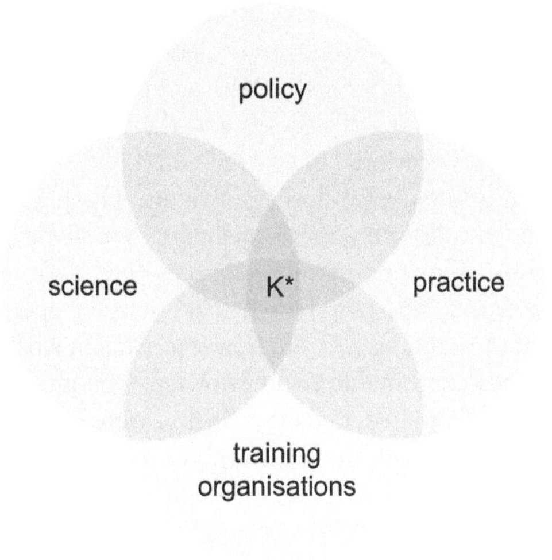

Figure 11.2 A systems view of knowledge dissemination
Source: Adapted from Shaxson et al., 2012

participatory sustainable agricultural development vocational learning environments as the demand for responding to ongoing ecological degradation continues apace (Masara, 2010; Mukute, 2010; Lotz-Sisitka, 2012; Lotz-Sisitka et al. 2016; Pesanayi, 2019). Apart from the cognitive gains that accrue from learning, social learning, and especially transformative social learning, is also interested in the actual social and practical transformations that result from learning, and how such transformation can be facilitated through experiential and change oriented learning interactions and knowledge exchange (cf. Chapter 10). Such processes of learning can also be mobilised as a mediating approach in forming regional social ecosystemic approaches to skills development (Hodgson & Spours, 2016, 2018; cf. Chapter 8). As noted in Chapter 8, establishing regional skills ecosystemic approaches requires *mediation*, but as yet, not much has been said about this mediation, and how it is to be constituted, an issue which we address in this chapter.

Aligned to these understandings of learning and change is a growing body of knowledge on the issue of knowledge dissemination which also seeks to address the knowledge flow problem. Scientific organisations globally are confronting problems of knowledge dissemination and uptake, for example, the UK Department for International Development (DFID) launched a 'Knowledge-in-use' programme (2006–2012) (DFID, 2016) and there is burgeoning practitioner literature on information intermediation, knowledge translation and knowledge brokerage (Blackmore et al., 2011; Shaxson et al., 2012). Likewise, there is increasing academic literature on transdisciplinary practice (Regeer & Bunders, 2009; Colvin, Everard, & Chuimbuya, 2011) focusing on mediation of knowledge in practice through knowledge co-production processes involving researchers and practitioners working together on knowledge and practice problems or innovations. This literature, however, tends to focus on 'knowledge flow' between researcher and practitioner, and tends to exclude educators or educational institutions, which Hodgson and Spours (2018) argued are important mediators in the building of regional social skills ecosystems that seek to join up work, living and learning in regional vocational education contexts.

Shaxson et al. (2012) defined a knowledge flow continuum which includes a diversity of knowledge mediation roles such as *information intermediaries*, who help people to access information from one or more sources (e.g. radio, local resource centres), *knowledge translators*, who help people make sense of and apply educational knowledge (e.g. agricultural colleges), *knowledge mediators*, who work with others to use knowledge in decision making and knowledge co-production processes through participatory and applied practices (e.g. extension officers, non-governmental organization (NGO) fieldworkers) and *knowledge innovators*, who work with existing knowledge in contexts of practice to facilitate new knowledge production and social innovation (e.g. action researchers, formative intervention researchers, communities of practice), the latter being more systemic than the former.

Methodology

We addressed the knowledge flow problem via a formative intervention case study that sought to use generative co-engaged approaches to understanding and supporting the uptake and use of rainwater harvesting knowledge resources that had been developed by the Water Research Commission (WRC). These knowledge resources were available for use by knowledge intermediaries (radio stations), translators (e.g. agricultural colleges), mediators (NGOs and extension services) and knowledge brokers (local economic development (LED) officers), but were not being used. The formative intervention case study, which started in 2014, was called 'Amanzi [Water] for Food' as a strategy to make the core concept being mediated by the materials clearer to those who were most interested in the knowledge (see amanziforfood.co.za).

We initially used the Shaxon et al. (2012) K* framework to help conceptualise a more systemic approach to knowledge flow that did not make the same mistakes as the RDDA approach. This helped to offer a wider view of the vocational agricultural learning system that went beyond the concept of Vocational Education and Training being confined to college walls, knowledge translation and accredited training or information sharing via media. Instead we adopted a transformative social learning mediation approach involving multi-actors in the local agricultural learning system (cf. Chapter 10), best described (in hindsight) within a social ecosystemic skills development orientation (cf. Hodgson & Spours, 2016, 2018; cf. Chapter 8) that was also place-based, with the first learning network being established in the Nkonkhobe Municipality in the rural Eastern Cape. This was necessary because few of the agricultural education and training institutions that we engaged with initially were focusing on the needs of smallholder farmers, or on rainwater harvesting and conservation practices. Extension officers were not well equipped to support farmers with this knowledge as it had traditionally been absent in their own training, and yet the Eastern Cape farmers were desperate for knowledge of how to harvest and store and thus bring water into the local food production system.

Unlike the RDDA model, a systems view sees knowledge 'flow' at the intersection or 'nexus' of different life worlds, in 'constant flow' and interaction (cf. Figure 11.2). It recognises that knowledge flows are influenced by structural, organisational, individual and system challenges (cf. the critical realist underlabouring shared in Chapter 13), including knowledge mediation challenges exhibited in different contexts. For example, to successfully apply or work with the WRC research-based knowledge resources in agricultural colleges, there is need to analyse and understand the structural, individual, organisational and wider institutional and system challenges (cf. Bhaskar's laminated system in Chapter 13) faced by its managers, lecturers and curriculum developers in order to develop strategies that will enhance knowledge flow, mediation or uptake and use of new knowledge in this context. There is also need to consider the links between agricultural colleges and other actors in the agricultural learning system to expand the concept of knowledge dissemination. Therefore, *K** (knowledge dissemination and use approaches) in the context of the Amanzi for Food research project referred to:

> a collective term for the set of functions and processes at the various interfaces between science, policy, practice and training organisations that improve the sharing of knowledge, its application, uptake and value in the pursuit of progress towards productive water use for food crop production amongst small-scale farmers and household food producers in South Africa.
>
> (Definition adapted from Shaxson et al., 2012, cited in Lotz-Sisitka et al., 2016, p. 33)

Our study further extended the systems approach to knowledge dissemination by drawing on learning network theory, which we used to consider the processes of establishing a learning network. According to Cousin and Deepwell (2005), a *learning network* is a group of people coming together to learn about certain topics of interest to all members. These networked groups are potential platforms for learning to occur (Weaver, 2016) as members bring their own experiences and competences from their diverse backgrounds, creating opportunities for collaborative learning and engaged practice (Wenger et al., 2011). According to Weaver (2016, p. 32), a network can be a very effective learning resource when it is designed to sustain, accept and utilise learner differences, such as their diverse competencies (Cousin & Deepwell, 2005) and individuals act as nodes and encourage information flows in the broader network (Wenger et al., 2011). A case study in Hobeni, Eastern Cape showed that agricultural resources and information can be disseminated effectively through facilitating and building effective networks (Fay, 2010), where partners in these networks share a great deal of knowledge even with minimal interaction (Weaver, 2016). Weaver (2016). This suggested that a successful learning network requires its participants to be flexible, responsive and continually learn from one another. Most importantly, the partners should strike a balance between inside (experiential and internal knowledge) and outside (external research knowledge) knowledge to form successful collaborations (Lieberman, 2000).

The learning network also required a different approach to knowledge mediation, as the traditional concept of a one-way knowledge flow or even Bernstein's (1990) concept of 'recontextualisation' via the pedagogical device appeared to be inadequate for constituting the kind of complex social ecosystemic knowledge flow that was needed in the local agricultural learning system. We have named this an *'iterative'* approach to knowledge mediation. To inform this approach, we probed histories of knowledge mediation in the water sector, noting that knowledge mediation involving the 'flow' of water science and practice knowledge into educational and community settings has been a problem in South Africa (Burt & Berold, 2012; Burt et al., 2014). Burt and Berold (2012) and Burt et al., (2014) noted that, most often, knowledge was not shared or mediated in a way that encouraged individuals to question their practice or consider how to adapt it. The presentation of 'factual' packaged information was insufficient, and it was often inaccessible to non-specialists (ibid.). They reported that learning resources tend to work more effectively when learners are engaged with water issues in their local context as they experience them (Burt & Berold, 2012). Burt et al. (2014) and Rivers (2017), drawing on Vygotskian (1978) insights into mediation and meaning making, recommended mediation of water via models of co-creation, where both knowledge and the media are developed collaboratively with researchers, 'mediators' (trainers and facilitators) and users all working together around practices or shared activity (Lotz-Sisitka et al., 2016), which led us to a focus on practice (discussed below) as central to the mediation processes. Burt et al. (2014) noted further that "knowledge mediation requires that careful attention be given to a range of contextual factors that are social-ecological or socio-material, historical-educational, linguistic and cultural", which reminded us to undertake ongoing contextual analysis within a critical realist laminated systemic framework (Lotz-Sisitka, Meterlekamp, & Ramsarup, 2019).

To deepen our views on learning and mediation around practices and activity, we drew on Engeström's post-Vygotskian (2007; 2014) research which expanded Vygotsky, Leont'iev and Luria's work on learning and activity, explaining that knowledge flow and learning are *culturally historical. i.e. situationally mediated processes around shared human activity*, in our case, the activity of rainwater harvesting and conservation for smallholder farming and household food production. Vygotsky (1978) claimed that humans use tools developed from a culture (e.g. speech and/writing) to mediate their own learning and social environments. For example, concepts associated with rainwater harvesting and conservation (RWH&C) such as 'saaidamme' [meaning medium sized water harvesting dams], and/or practices or artefacts (e.g. mulching). When these tools are internalised, the higher order thinking skills are developed (Vygotsky, 1978) and a person's learning can be expanded. Engeström (2007; 2014) argued that activity can be expanded via collective, multi-voiced learning mediated by formative interventionist researchers in boundary crossing change laboratories where multiple actors from interacting activity systems are present. In our case, the activity systems were the agricultural college activity system, the farmers activity system, the extension activity system, the media activity system and the local development activity systems, all of whom shared an interest in developing rainwater harvesting and conservation practices in support of smallholder farmers' food production activity in the municipal area.

In the Vygotskian theory of social development, social interaction and social learning *precedes development (i.e. learning leads to development)*. This is significant for the notion of knowledge flows because for knowledge 'uptake' to succeed mediated learning should happen. Thus, knowledge is not simply absorbed or 'taken up' without *the process of meaning-making*, commonly elided or neglected in RDDA approaches and assumptions. This also explains why this type of research *needs to be framed as generative, co-engaged research*. In this case, participants acted and interacted in shared experiences and further mediations of knowledge and meaning-making processes, which often occurred in non-linear ways, hence we used the term *iterative mediation processes*.

Practice analysis, contextual profiling and navigation tool for supporting knowledge access and mediation in context

The preparation for knowledge mediation from the WRC materials project design followed different phases. First, the key RWH&C practices in the WRC materials within a social learning network were identified for use at different levels of small-scale farming and household food production. We identified 16 different rainwater harvesting and conservation practices that were presented in the WRC materials, and classified these according to type of practice, relevance to the scale of farming activity, the main purpose of the practice, factors to consider with regard to the practice and noted where farmers and mediators could find information on the practice in two sets of WRC materials (described in Lotz-Sisitka et al., 2016). This analysis informed the development of a *navigation tool* used for accessing knowledge in the WRC materials in the learning network. The construction of the navigation tool is shown in Table 11.1 with the full navigation tool available on the website www.amanziforfood.co.za.

Contextual profiling and knowledge mediation processes

Doing the practice analysis from the materials as outlined in Table 11.1 above was, however, not enough. We also needed to know what practices farmers were using (or not). We therefore conducted a contextual profiling process concurrently to understand the water for food problems faced by smallholder farmers, and how agricultural colleges, universities and extension services were implicated in agricultural water knowledge flow along the agricultural learning system. Concomitant with the practices review and informing it, the contextual profiling was also aligned with the knowledge mediation tools (primarily the navigation tool outlined in Table 11.1). Contextual profiling involved interviews with key informants from smallholder farmers and household food producers in the learning network site; agricultural lecturers from agricultural training institutes, universities and further education and training (FET) colleges; local economic development facilitators; extension advisors, other farmers, farmers' association members, research institutions and other actors in the local agricultural learning system (e.g. NGOs). Focus group discussions and participant observations were also conducted during the contextual profiling (Lotz-Sisitka et al., 2019).

The contextual profile revealed four key outcomes around agricultural water knowledge: knowledge flow, learning support materials, information exchange and dissemination media and the politics of knowledge regarding what was considered important (Pesanayi, 2019). First, the three agricultural colleges involved in the WRC research processes did not significantly use the learning support materials that were emailed to them by the WRC in their curriculum as noted above, confirming the knowledge flow problem. Additionally, knowledge of rainwater harvesting practices was very weak or non-existent amongst agriculture lecturers in agricultural colleges and universities, because this was lacking in their teaching curricula (Pesanayi, 2019), and only a few people among the college lecturers, smallholder farmers and extension advisors were familiar with more than three RWH&C practices, showing that knowledge flow from the WRC materials to agricultural colleges and farmers in the sites was virtually non-existent.

Additionally, the extension advisors did not promote the little RWH&C knowledge acquired from previous Agricultural Research Council projects in the same area in the Eastern Cape Province (Pesanayi, 2019). This could be connected to the lack of RWH&C training in agricultural colleges and university curricula where these extension advisors had been trained. This omission or absence could also be due to the politics of knowledge whereby conventional agricultural water science knowledge is privileged over other agricultural water knowledge of small-scale or rudimentary application, and we found that smallholder farmers were not included in agricultural college curricula, due to the historical privileging of large scale farming, despite government policy promoting support for smallholder farmers. Also, there were very poor connections among farmers and between the various stakeholders through social media. However, there were strong links to agricultural programmes on community radio.

The contextual profiling work helped to raise these concerns in the social context, and there was willingness amongst farmers and other actors to form a learning network, as all realised that the water issues facing the smallholder farmers were serious and were affecting local food security. The formation of a learning network required mediation for the emergence of new, meaningful learning and human activity that would build knowledge of RWH&C amongst the participants. The contextual profiling and development of the navigation tool offered important starting points for mediation of co-engaged social learning in the learning network, where participants managed to use the WRC materials in different ways throughout the process.

Table 11.1 Sample from the navigation tool, showing the information shared to provide access to knowledge on the different RWH&C practices

Storing rainwater				
Practice	*Purpose(s)*	*Scales/Styles*	*Other factors*	*WRC materials: Text (T), Case Studies (CS), Handouts (H), Videos (V)*
Dams	Harvest and store	Mostly 3, but also 2 and possibly	Medium technology, specialised equipment, medium skills and understandings, medium to high cost (depending on scale)	WH&C (**T**: Pp.18), AWHGS (Vol.2, Part2, **T**: Pp.5–22, 5–85, 5-87-89, **CS**: Pp.5–87)
Matamo/ipitsi (homestead ponds)	Harvest and store	Mostly 1, possibly 2	Low technology, basic equipment, low skills and understandings, low cost, low maintenance	WH&C (**T**: Pp.18), AWHGS (**T**: Vol.2, Part2, Pp.5–26), IWHCP (**CS**: Pp.26–29 and **V**)
Underground tanks	Harvest and store	1, possibly 2	Medium technology, specialised materials and equipment, medium skills and understandings, medium cost, medium maintenance	AWHGS (**T**: Vol.2, Part2, Pp5–36 and **H**: Vol.2, Part2, Ch.5, H1, Pp.11–12), **V**: DVD1
Roof tanks	Harvest and store	1	Medium technology, medium skills and understandings, medium cost	WH&C (**T**: Pp.158), AWHGS (**T**: Vol.2, Part2, Pp5–83 and 5–84 and **H**: Vol.2, Part2, Chap.5, H1, Pp.9–11), **V**: DVD1

- Scale 1: Household/homestead (Garden, Umzi) < 1 ha – associated with Style 1, Food farmers (primarily subsistence, growing for own use)
- Scale 2: Field 1–2 ha – associated with Style 2, Employers (subsistence plus some low-level, local commercial production)
- Scale 3: Farm >2 ha – associated with Style 3, Profit-makers (primarily for local, regional and national commercial production)
- WH&C = Water Harvesting and Conservation Materials
- AWHGS = Agricultural Water Use in Homestead Gardening Systems materials

Generative development of the five iterative mediation processes supporting social learning network formation and co-learning in the learning network

The Eastern Cape Learning Network has been in existence since 2014. *Five significant mediation processes* were generatively identified, developed and used to support the ongoing co-learning in the Eastern Cape Learning network, catalysing a further two learning networks in the Amanzi for Food programme which started after 2017 in North West and Mpumalanga Provinces. These five mediation processes were *iteratively* developed and described during the process of forming the Eastern Cape learning network, and have informed similar learning networks in Mpumalanga and North West. The iteratively developed mediation processes were:

- Co-engaged need state analysis and learning network formation
- Boundary crossing change laboratories
- Course activated engagement around new knowledge of RWH&C practices
- Productive demonstration site development
- Expansion of knowledge engagement via use of social media and radio.

Iterative mediation process 1: Co-engaged need state analysis and formation of a learning network

The Shaxon et al. (2012) framework allowed us initially to identify the following knowledge partners in the vocational agricultural learning system: agricultural colleges, universities, technical and vocational education and training (TVET) institutions and/or private training providers; government departments, extension services and field-based change agents/fieldwork staff in the local economic development office as well as NGOs who were working on the object of agricultural water use practices for smallholder farmers and household food producers; social media partners such as community radio stations, and locally used social media platforms including the short messaging system (SMS) application WhatsApp. This informed the scope of our contextual profiling, where a similar range of actors were engaged (see above) along with the issues they were experiencing around RWH&C knowledge flow and practice.

A series of workshops were conducted to discuss issues related to food and water that culminated in an agreement to work together on the common problems identified. This initiated the first learning network meeting, where stakeholders agreed to establish the Imvotho Bubomi [Water is Life] Learning Network in the Eastern Cape (Weaver, 2016; Pesanayi, 2019). The local agricultural college, Fort Cox Agriculture and Forestry Training Institute, became a 'coordinating heart' as a local learning centre/TVET college. Importantly, and as also recommended in the social ecosystemic skills model of Hogson and Spours (2016, 2018), the local vocational college/agricultural universities were requested to 'host' the learning networks, given their mandate for supporting agricultural learning which also includes the new mandate of supporting smallholder farmers' learning and training extension officers to do the same. The learning networks members indicated that they had interconnected knowledge gaps with no ready solutions.

The research team borrowed from and followed guidance from cultural historical activity theory to implement an approach of using boundary crossing change laboratory workshops (Engeström & Sannino, 2010) to co-engage participants in analysing their need state(s) and how these were reciprocated. One smallholder farmer (21July 2016) expressed the reciprocity as follows:

> We are crying for that as farmers... there must be something that is relevant from these institutions (referring to ATIs) to what we are doing. That is what we were crying for from that workshop (held May 2016) and we are still crying for that.

Iterative mediation process 2: Boundary crossing change laboratories

Boundary crossing change laboratories (Engeström & Sannino, 2010) were set up on three sites in the Eastern Cape Learning Network. Members from diverse activity systems participated actively using data from the need state analysis (Pesanayi, 2019), following the Expansive Learning Cycle (Figure 11.3). Participants moved together through a series of learning actions as outlined in Figure 11.3.

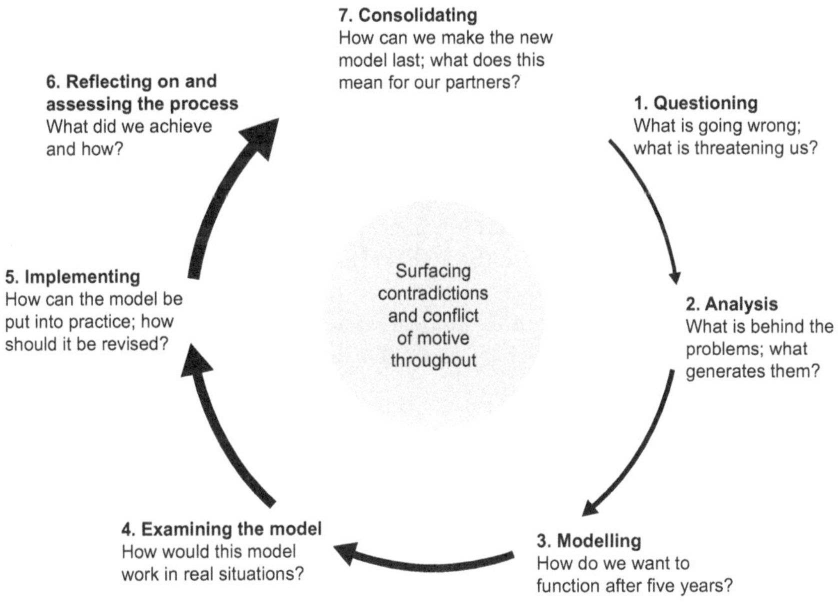

Figure 11.3 Expansive learning process followed in change laboratories
Source: Adapted from Engeström & Sannino, 2010

In this process, participants deliberated on water for food problems and curriculum gaps. Various new solutions to address the need state emerged with '*gelesha*' being an interesting but little-known indigenous knowledge practice of ripping the soil to conserve rainwater. Farmers and lecturers previously unfamiliar with *gelesha* developed an interest in the practice because of its local cultural appeal. Stories of innovation and WRC rainwater harvesting materials were used to stimulate expansive learning of new solutions and the development of new activity. This resulted in boundary crossing engagement across agricultural learning activity systems represented by different stakeholders (cf. Pesanayi, 2019 for full details). Research participants requested training to learn more about RWH&C practices.

Iterative mediation process 3: Course activated knowledge engagement

A university-accredited course was designed to activate knowledge engagement around new knowledge of RWH&C practices into applied change projects, using the navigation tool to guide participants to a wider range of potential RWH&C practices. This was designed to accommodate both academically advanced persons and those at a more basic level, thus helping with selecting, examining, refining and prioritising emerging solutions. A co-engaged participatory learning course was pitched as a training-of-trainers (ToT) to encourage participants to share their knowledge with others. A five-stage co-engaged participatory course (work-together, work away), was designed to develop change projects in curriculum innovation (in agricultural colleges) and farming practice innovation (with smallholder farmers) over a nine-month period (Weaver, 2016). The course content was designed to fill the knowledge gaps identified in the aforementioned needs analysis stage, and to accommodate the emergence of new solutions independently developed by research participants. The course enabled engagement with practical knowledge from WRC materials connected to the history of the object of agricultural water. It also facilitated engagement with diverse methods of alternative water for food and rainwater harvesting practices as found in the WRC materials, and captured in the 16 RWH&C practices outlined in the navigation tool. A key focus of the course process was to develop a change project that included co- development of an appropriate productive demonstration site on RWH&C, one that was socially acceptable, technically and environmentally feasible and economically viable.

Iterative mediation process 4: Productive demonstration site development

Encouraged via the course, participants in the learning network formed multi-stakeholder teams comprising of smallholder farmers, agricultural college/ university lecturers and extension advisors. In the Eastern Cape, four heterogeneous productive demonstration sites were selected and worked on collectively.

These included a cooperative garden in Lloyd Village, Fort Cox College (now an agriculture and forestry training institute) farm and two family food gardens in Keiskammahoek (homestead garden) and Matole Basin (cropland farm and homestead garden) (Pesanayi, 2019). Participants selected different rainwater harvesting practices, namely farm ponds, a farm dam connected to a raised tank, drip irrigation over a mulched garden and mulched circular beds, respectively. *Productive demonstration sites* can be described as *on*-farm or *off*-farm physical structures that *capture*, divert, store, *sink* and conserve rainwater for future productive use by crops and pasture/livestock (Pesanayi, 2019; Lotz-Sisitka et al., 2016). These sites were developed through collaborative work efforts called *ilima* in isiXhosa. The main test for new learning was the actual use of the demonstration sites for productive agriculture. Important for the knowledge flow problem, we found participants actively using and drawing on the formerly 'unused' WRC materials to inform their collective practice.

Family garden plots continued to be used by smallholder farmers to improve their production and they began to regularly host student training by the local agricultural college and Further Education and Training (FET) college lecturers around the concept of the productive demonstration site (i.e. the activity of RWH&C). The farmers acted as co-facilitators of the training which made the students consider the sites as authentic learning sites. This transformed the hitherto overemphasis on theory in the college, where no practices of RWH&C were found in the contextual profiling phase. The lecturer who was part of the team working at a productive demonstration site in Keiskammahoek area described this scenario as follows:

> Wait till you listen to the video. The way the garden looked and how the technique is working! Also, the money she is making on that piece of land. She was excellent in her teaching. I think the teaching place, who delivers the lesson, plays a significant role in the learning. In a way it looks as if when they learn in class and through a lecturer is all about assessment but with the farmer base it is a real-life application.
>
> (Agricultural college lecturer, WhatsApp communication, 14 December, 2016)

The cooperative garden was the least successful productive demonstration site mainly because of the advanced age of the participating farmers. They felt they had little energy to maintain their farm ponds, especially beyond the first season despite having enjoyed success from the RWH&C in that growing season. However, youth farmers within the area took over these productive demonstration sites, also teaching younger scholars from the local school, thus showing the expansive learning potential of the productive demonstration site as a mediation tool. The utility of these sites was aligned to the notion that "humans act collectively, learn by doing, and communicate in and via their actions" (Vygotsky, 1978).

Iterative mediation process 5: Expansion of knowledge engagement via use of social media and community radio

Knowledge engagement continued to be mediated through social media and community radio which was established together with the research project team. Most participants had smartphones to communicate through WhatsApp, therefore a WhatsApp group was formed for the Imvotho Bubomi Learning Network in May 2014 (Weaver, 2016). The group has been operational and growing throughout the project duration. WhatsApp has proved to be a powerful knowledge mediation and information exchange tool, complementing learning network meetings and *ilima* activities. One drawback of these groups was the exclusion of those few members with no WhatsApp capability. The success of the WhatsApp group learning network gave birth to sub-WhatsApp groups that included community radio.

A local community radio, Forte FM, was invited to be part of the Imvotho Bubomi Learning Network from the time of its inception which enabled the securing of a convenient monthly slot. The learning network would discuss matters of concern and share wise RWH&C practices that had emerged in their learning network, which they had tried and tested (Lupele, 2017). New stories of significant change and innovation were documented on Facebook and website blogs (www.amanziforfood.co.za). At the centre of these processes was the co-engaged uptake and use of the valuable RWH&C knowledge formerly unused in the WRC materials.

Conclusion

This chapter has argued for a systems approach to knowledge flow, and for a generative formative interventionist research approach to the mediation of knowledge flow in social ecosystemic skills vocational learning contexts. It has proposed the learning network as an appropriate learning structure within local small-scale and household food production systems. Furthermore, it has argued for the need to give attention to a range of iteratively constituted mediation processes that are necessary for a systems approach to knowledge flow, highlighting social learning and innovation. These were briefly shared in broad terms with some of the emergent results of the processes beginning to show up (in-depth reports on these are available in Lotz-Sisitka et al., 2016; 2019; Weaver, 2016; Lupele, 2017, Sithole, 2017; Pesanayi, 2019). The chapter noted the importance of preparing well for mediation processes, not in terms of preparing 'pre-determined lessons' but rather in terms of coming to understand the issues associated with knowledge flow and mediation, as well as understanding the context well while also knowing the valuable practices and knowledge resources that can be mobilised in the learning setting and network.

Results captured in the Amanzi for Food programme to date (ibid.) show that the established learning networks continue to expand and deepen with expansions of the mediation processes. Also, smallholder farmers have gradually been placed at the centre of knowledge production and application, to whom agricultural college educators could relate their curricula. There is also emerging evidence that agricultural educators in colleges and universities are innovating their curricula and programmes to align with the realities of the smallholder farmer and with more sustainable agricultural practices. That is, they are including what they learned via the course, media, learning network and productive demonstration sites in their teaching, and are also introducing and expanding the content and practical teaching of RWH&C to other areas of sustainable agriculture in the service of smallholder farmers. The learning network, supported by the five iterative mediation processes, strongly developed the horizontal connectivities that Hodgson and Spours (2018) refer to in their social ecosystemic skills development model, and the use of the WRC materials is a good example of 'facilitating verticalities' (cf. Chapter 8), both necessary for strengthening vocational learning systems. The productive demonstration sites were significant for addressing the problem of a loss of practice in vocational educational curricula, and also strengthened the possibility of a more community-engaged vocational education curriculum and systemic learning process.

Acknowledgements

In-depth reports on the processes described here are available in Lotz-Sisitka et al. (2016; 2019), Weaver (2016), Lupele (2017), Sithole (2017) and Pesanayi (2019), indicating that this programme has been a large team effort over a period of time. Further acknowledgements are contained in the reports but the farmers and colleges who participated in the research are especially acknowledged. The research leadership provided for this project by the late Dr Tichaona Pesanayi, captured in his PhD (2019), is acknowledged and appreciated. Tichaona sadly passed away in the very early stages of work on this chapter. The authors would also like to acknowledge and thank the Water Research Commission of South Africa for funding the Amanzi [Water] for Food project (cf. Water Research Commission Report No. 2277/1/16), and the National Research Foundation (NRF) and the Department of Science and Technology (DST) for their support of the SARChI Chair in Global Change and Social Learning systems whose core research funding also supported the writing of this chapter.

Note

1 The paper was initially conceptualized with Tichaona Pesanayi, he drafted some of the early rough text on the mediation processes, but sadly passed away shortly after we started working on the paper.

References

Backeberg, G.R., & Sanewe, A.J. (2010). *Towards productive water use and household food security in South Africa.* Paper presented at the 6th Regional Conference of ICID held at Yogakharta, Indonesia, 10–16 October 2010.

Bernstein, B. (1990). *The structuring of pedagogic discourse, Volume, IV: Class, codes and control.* London: Routledge.

Blackmore, C., Chabay, I., Collins, K., Gutscher, H., Lotz-Sisitka, H., McCauley, S., Niles, D., Pfeiffer, E., Ritz, C., Schmidt, F. Schreurs, M., Siebenhuner, B., Tabara, D., & Van Eijndhoven, J. (2011). *Knowledge, learning, and societal change: Finding paths to a sustainable future. Science plan for a cross-cutting core project of the International Human Dimensions Programme on Global Environmental Change* (IHDP). Bonn: IHDP.

Burt, J., & Berold, R. (2012). *Investigating water knowledge flow to communities.* Water Research Commission Research Report No KV 288/11. Pretoria: Water Research Commission.

Burt, J., Lotz-Sisitka, H., Rivers, N., Berold, R., Ntshudu, M., Wigley, T., Stanford, M., Jenkin T., Buzani, M., & Kruger, E. (2014). *The role of knowledge in a democratic society: Investigations into mediation and change-oriented learning in water management practices.* Water Research Commission Research Report No. 2074/1/13. Pretoria: Water Research Commission.

Colvin, J., Everard, M., & Chuimbuya, S. (2011). *Researching transdisciplinary approaches in the Mvoti sub-catchment as a means to progress an ecosystem services approach to poverty alleviation in South Africa.* Water Research Commission Research Report K5/2033/02. Pretoria: Water Research Commission.

Cousin, G., & Deepwelli, F. (2005). Designs for network learning: A communities of practice perspective. *Studies in Higher Education*, 30(1), 57–66.

Denison, J., Smulders, H., Kruger, E., Ndingi, H., & Botha, M. (2011). Development of a comprehensive learning package for education and training on the application of water-harvesting and conservation. Water Research Commission Report No. TT/492/11. Pretoria: Water Research Commission.

Department for International Development (DFID). (2016). Research uptake. A guide for DFID-funded research programmes. London: DFID.

Engeström, Y. (2007). Putting activity theory to work. In H. Daniels, M. Cole & J.V. Wertsch (Eds.), *The Cambridge companion to Vygotsky*, pp. 363–382. Cambridge: Cambridge University Press.

Engeström, Y. (2014). *Learning by expanding.* Cambridge: Cambridge University Press.

Engeström, Y., & Sannino. A. (2010). Studies of expansive learning: Foundations, findings and future challenges. *Educational Research Review* 5(1), 1–24.

Fay, D. (2010). Cultivators in action, Siyazondla inaction? In *Trends and potential in homestead cultivation*, pp. 1–29. Oakland, CA: University of California.

Hodgson, A., & Spours, K. (2016). The evolution of social ecosystem thinking: Its relevance for education, economic development and localities: A stimulus paper. London: Centre for Post-14 Education and Work, UCL Institute of Education.

Hodgson, A., & Spours, K. (2018). *A social ecosystem model: Conceptualising and connecting working, living and learning in London's New East.* The East London Vocational Education and Training: Innovation through Partnership Programme (ELVET). ELVET Research Briefing No. 3, March 2018.

Labradarios, D., Mchiza, Z.J., Steyn, N.P., Gericke, G., Maunder, E.M., Davids, Y.D., & Parker, W. (2011). Food security in South Africa: A review of national surveys. *Bulletin of the World Health Organization* 89, 891–899. doi: 10.2471/BLT.11.089243

Lieberman. A. (2000). Networks as learning communities: Shaping the future of teacher development. *Journal of Teacher Education*, *51*(3), 221–227.

Lotz, H.B. (1995). *The development of environmental education resource materials for junior primary education through teacher participation*. Unpublished PhD thesis, University of Stellenbosch, South Africa.

Lotz-Sisitka, H.B. (Ed.) (2012). *(Re)Views on social learning literature: A monograph for social learning researchers in natural resources management and environmental education*. Grahamstown and Howick, South Africa: Environmental Learning Research Centre, Rhodes University/EEASA/SADC REEP.

Lotz-Sisitka, H., Pesanayi, T., Weaver, K., Lupele, C., Sisitka, L., Denison, J., & Van Staden, W. (2016). Water use and food security: Knowledge dissemination and use in agricultural colleges and local learning networks for home food gardening and small scale agriculture. Volume 1: Research and development report. Water Research Commission Report No. 2277/1/16. Pretoria: Water Research Commission.

Lotz-Sisitka, H.B., Meterlekamp, L., & Ramsarup, P. (2019). *Contextual profiling: A laminated systems approach*. Unpublished paper, Environmental Learning Research Centre, Rhodes University, Grahamstown, South Africa.

Lupele, C. (2017). *A review of the development and enacting of a radio programming process in expanding an emerging community of practice and enhanced social learning*. Unpublished Master's thesis, Rhodes University, Grahamstown, South Africa.

Masara, C. (2010). Social learning processes and nature-culture relations of commercial beekeeping practices as small and medium enterprise development in Zimbabwe. *Southern African Journal of Environmental Education*, 27, 10–20.

Mukute, M. (2010). *Exploring and expanding learning in sustainable agriculture practices in southern Africa*. Unpublished PhD thesis, Rhodes University, Grahamstown, South Africa.

Pesanayi, V.T. (2019). *Boundary-crossing expansive learning across agricultural learning systems and networks in southern Africa*. Unpublished PhD thesis, Rhodes University, Grahamstown, South Africa.

Rivers, N. (2017). *The mediating processes within social learning: Women's food and water security practices in the rural Eastern Cape*. Unpublished PhD thesis, Rhodes University, Grahamstown, South Africa.

Regeer, B., & Bunders. J. (2009). *Knowledge co-creation: Interaction between science and society – a transdisciplinary approach to complex societal issues*. Amsterdam: RMNO.

Robottom, I. (1991). Technocratic environmental education. A critique and some alternatives. *Journal of Experiential Education*, *14* (1), 20–26.

Shaxson, l., Bielak, A., Ahmed, I., Brien, D., Conant, B., Fisher, C., Gwyn, E., Klerkx, l., Middleton, A., Morton, S., Pant, l., & Phipps, D. (2012). *Expanding our understanding of K**. In UNU-INWEH K* 2012 Conference, Hamilton, ON, Canada, April 2012.

Sithole, P. (2017). *Investigating the role of extension officers in supporting learning of rainwater harvesting practices amongst rural smallholder farmers in Nkonkobe Local Municipality, Eastern Cape*. Unpublished Master's thesis, Rhodes University, Grahamstown, South Africa.

Vink, N., & Van Rooyen. J. (2009). *The economic performance of agriculture in South Africa since 1994: Implications for food security*. Pretoria: Development Planning Division, Development Bank of Southern Africa.

Vygotsky, L. (1978). *Mind in society: The development of higher psychological processes*. Cambridge, MA: Harvard University Press.

Weaver, K. (2016). *Exploring the course-led development of a learning network as a community of practice around a shared interest of rainwater harvesting and conservation agricultural practices: A case study in the Amathole District in the Eastern Cape, South Africa*. Unpublished Master's thesis, Rhodes University, Grahamstown, South Africa.

Wenger. E., Trayner, B., & De Laat, M. (2011). *Promoting and assessing value creation in communities and networks: A conceptual framework*. Heerlen, Netherlands: Ruud de Moor Centrum.

World Food Programme (WFP). (2016). The year in review 2016. Retrieved from https://docs.wfp.org/api/documents/WFP-0000019183/download/?_ga=2.237156480.1439052066.1570531162-2124015837.1570531162

12 Greening occupations and green skills analysis

Presha Ramsarup

Researching the nature of green work and the greening of jobs

Within South Africa, the environment and sustainability is largely a cross-cutting issue; hence provisioning for the sector should ideally be across various skills entities. However there are several challenges with regard to provisioning for green jobs. Planning and provisioning for environmental skills development in South Africa has been '*inadequate, ad hoc, fragmented, reactive and inefficient*' (Department of Environmental Affairs [DEA], 2010, p. 5). The Environmental Sector Skills Plan study estimated shortages of over 800 environment-related scientists and 1 500 environment-related technical skills in the public sector, affecting service delivery and implementation of core national legislation. Despite a dire need to up-skill staff, environmental agencies have been unable to effectively source Sector Education and Training Authority (SETA) funding (DEA, 2010). Employers tend to provide generic training at levels too low to address identified skills needs and line function budgets are often used to provide more relevant training (DEA, 2010), which reflects an inefficient use of funding at a broader systemic level.

This was compounded by newness, the virtual absence of environmental occupations in the national system of occupations and the realisation that the systems for collecting information on environmental occupations were also almost non-existent. This was a significant determining factor for allocation of resources for skills planning and development in South Africa, as the language of occupations and 'occupation' is the unit for allocation of resources. Absences such as this have knock-on effects on skills planning, national monitoring and evaluation (e.g. absence in the Labour Force Survey of Statistics South Africa (STATS SA)) and result in inadequate capacity and tools for understanding the sector and its changing nature, for example, how researchers had to calculate waste workers in municipalities for the Environmental Sector Skills Plan(ESSP) study.

In South Africa there have been a number of green job projections at large-scale aggregate levels, all focused on the numbers of green jobs that can be created in the short and long term (Maia & Giordano, 2011). They helped to illustrate that the green economy has implications for many sectors of the economy. They also gave us an idea of the size and scope of the environmental sector in South Africa. Furthermore, it is estimated that there will be a 45% increase in job creation in natural resource management in South Africa in the short term and 50% in the long term (Maia & Giordano, 2011). This narrative, however, remains an aspirational one; it misses some key aspects that were challenging to surfacing green skills, it did not speak to employers and provided little insight into how workplaces and actual work can transition into becoming greener or more sustainable. Thus, it remains disconnected from 'real' *jobs*. It does not translate into clear meaning for employers who often do not know what a green job is, nor do they understand where these jobs are or how to unlock them. Employers may lack a futures perspective and insight into future skills needs, particularly if their industries feature path dependence and lock-ins. In addition, these types of 'green job' statistics have a tendency to see green jobs as a 'one size fits all' and assume they are homogenous. There is a need for a more nuanced idea of what constitutes a green job. All these factors place constraints on traditional notions of demand or employer led demand and call for new and transformative skills methodologies.

Additionally, the important challenge for educational practitioners involved figuring out the educational and training implications. If jobs were changing, what would the knowledge needs be and how do education and training systems respond? Green job projections seldom answer educational questions. Thus newness, together with a vague and imprecise green jobs language, created a system restlessness that prompted a reactive and ad hoc educational response from vocational educational and training sectors that are still struggling to concretise their identity and scope.

Green: A contested terrain

The concept *green* is notoriously fuzzy; it is a socially constructed concept, intangible and often unobservable. This section highlights some of the complexities of this construct and how it creates problems when attempting to clearly articulate skill needs. Part of the fuzziness linked to green skills is their link to related terminology, the misconception related to the notion of green and the plethora of new and emerging green related concepts. The discussion below outlines how we worked to resolve some of these issues in the studies presented in this book.

A point of dissonance was that this discourse was new and separate from the concept of sustainable development. Georgeson, Maslin and Poessinouw (2017) provided a useful hierarchy of concepts and illustrated how most 'green' concepts are related and co-exist, dispelling the idea of conceptual inconsistency with sustainable development. They argued that a green new deal is a catalyst for green growth which makes a contribution to the green economy as a means of achieving sustainable development. This suggests that critical engagement with green economy discourses is central to sustainable development planning and implementation.

	Deep Green	Mid Green	Light Green
Mode	Pro-active	Integrative	Reactive
Scope	Long-term	Intermediate term	Short-term
Nature	Transforming	Reforming	Confirming
Objective	Redefine growth	"Ecologise" growth	Enhance growth
Operation	Rejectionist	Reinventionist	Accommodationist
Aim	Ecological sustainability	Ecological modernity	Sustainable development
Jobs	Preserving nature	Greening industry	Remedying ecological decline

Figure 12.1 Australian green jobs typology
Source: Goods, 2014

The ambiguity surrounding the notion of 'green' has led to different characterisations of green. There was little agreement in the literature or policy on what green jobs were, whether the term referred to new jobs or traditional jobs. In the literature we explored different typologies of green and how jobs were being classified. Green is clearly not an absolute notion. We investigated various frameworks for how green was conceptualised and utilised. The Australian framework presents a relatively comprehensive categorisation of Deep Green, Mid Green and Light Green (see Figure 12.1).

The International Labour Organisation (ILO) talks of 'direct' and 'indirect' green jobs, and uses the typology in Figure 12.2 to assess green and decent work.

From the descriptions we can see that an inclusive green economy is much more than an economic growth agenda; it presents an opportunity to advance both sustainability and social equity as functions of a stable and prosperous economic system. It is hence a pathway towards achieving the 2030 Agenda for Sustainable Development, supporting the eradication of poverty while safeguarding the ecological thresholds, which underpin human health, well-being and development.

In the contexts of social inequality, unemployment and ecological degradation as outlined in Chapter 1, it is also important to explore some of the varied contestations surrounding this emerging discourse at a more broad and philosophical level. Critiques of green economy discourses include that they are too closely aligned to current systems, do not consider potential limits to growth, are prone to oversimplification and misplaced optimism and the models used are questioned. All concepts relating to the green economy place the economic sphere at the centre of the debate on future viability. According to this view, we can only save the planet with the economy, not against it. So do all solutions

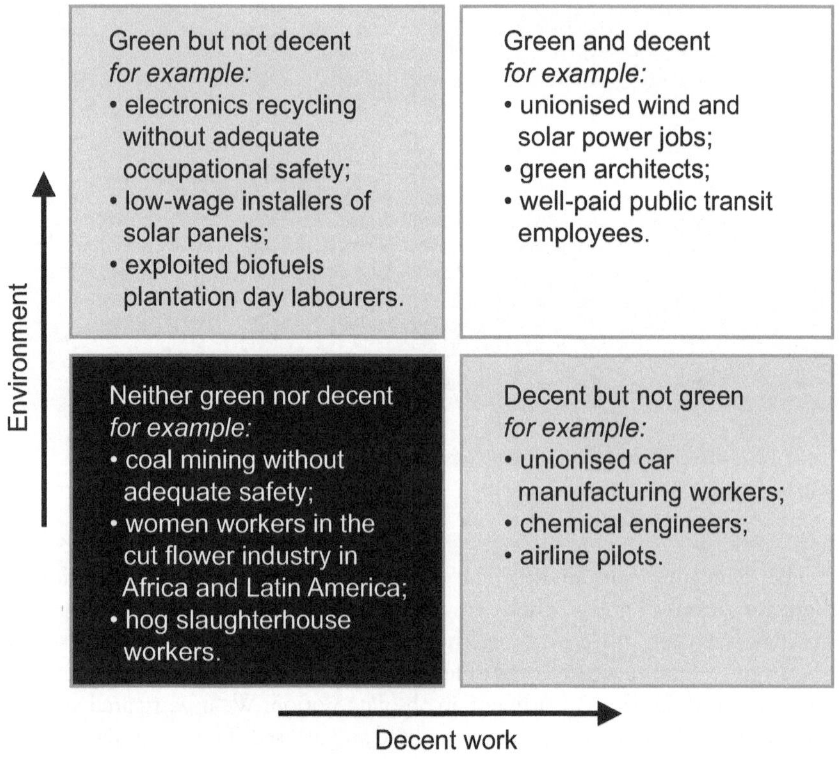

Figure 12.2 Framework for green and decent work
Source: ILO, 2016

revolve around *Homo economicus* once again (Unmüßig, Sachs, Fatheuer, & Unmüßig, 2012)?

Sociologist Cock (2012) has cautioned: "the 'green economy' is an empty signifier … everything depends on who claims it and gets to fill it with meaning" (p. 1). Cock argued that the environmental crisis "is a crisis arising from and perpetuated by the rule of capital and hence [is] incapable of resolution within the capitalist framework" (2012, p. 1). She claimed that capital is appropriating the climate crisis and promoting the further commodification and even financialisation of nature in the form of the 'green economy'.

In a similar vein, Gibbs and O'Neill (2014) have argued that the all-pervading United Nations Environment Programme (UNEP) definitions of green economy as well as the Global Green Deal, perpetuate business as usual, neo-liberal economics and growth and reliance on technology, rather than promoting a real paradigm shift that focuses on the holistic integration of social and environmental sustainability. They further argued that although this 'current greening' approach to growth and economics is green tinged, without constraints on consumption and demand we shall remain trapped in consumptive economies.

In supporting her thesis that the green economy discourse is a "wolf in sheep's clothing, the 'wolf' being green neo-liberal capitalism", which resonates with the Gibbs and O'Neill argument, Cock (2012, p. 2) raised two critiques that are useful: she argued that the emerging green discourse does not sufficiently engage with working class and justice issues, thus the evolving notions of justice are neglected or naïve. Secondly, the creation of green jobs is being viewed as something distinct and as an add-on to the 'real economy'. This, she argued, represents a compartmentalisation of the green economy discourse.

Adding his critical voice to support the arguments put forward by Cock, Death (2014) described the dominant green economy discourse in South Africa as a 'green growth' discourse. Green growth discourses present "green markets as an economic opportunity" (Death, 2014, p. 7) and the Green Economy Accord (EDD, 2011) presents four principles – opportunity, innovation, responsibility and partnership. Opportunity and innovation indeed suggest a green growth intent. Thus there is a growing community of critique outlining South Africa's emergent green economy response as promoting technological innovation and new markets and not addressing core concerns of social and environmental justice (Cock, 2012; Death, 2014).

Fergusson (2015) and Death provide two typologies that enable further critical engagement with these discourses. According to Death (2015) 'green transformation' centres around a complete transformation of the structure of the economy, green revolution, green growth and green resilience as discussed in Chapters 1 and 2. Ferguson's (2015) study of green economy discourses concludes that *green growth* discussions must be separated from *green economy* discussions. His study also identifies three categories of green economy discourse: *weak*, *transformational* and *strong*. Part of this categorisation includes assessing measurement and focusing on their relationship to gross domestic product (GDP). This contested terrain has significant implications for skills demand and supply research as has been illustrated across this book.

A just transition: moving forward

The emergent just transitions discourse has provided a way to move beyond the jobs versus environment argument. In its interactions with the environmental justice and climate justice movements 'Just Transition' has become a broad framing that supports an expanded scale of considerations across economic, social and environmental dimensions. Ward (2018) argues that Just Transition principles, processes and practices support both distributive and procedural justice as humanity grapples with issues of inequality, unemployment, climate change, resource scarcity and the need for rapid change.

Significant transitions are required by climate change and water shortages in South Africa. Similarly, global competition is driving a digital transition that has been called the Fourth Industrial Revolution. These changes are based on conscious decisions and, in many instances, policy, planning and financial support. This places a duty of care on those individuals and institutions participating in these decisions and transitions. An 'unjust transition' that leaves workers behind, abandons communities, deepens inequality, divides society and positions people against the planet all in the name of saving jobs and creating profit is not acceptable. Ward (ibid.) furthermore notes that a shallow or minimal transition focuses on preserving a few jobs with a narrow scale of focus in terms of current and future generations (time), local workers (geography), environmental issues and stakeholders. While this may respond to short-term interests and preserve some jobs, it is likely in the long run to make business and workers defenders of the indefensible.

These transitions require new knowledge and skills both within existing sectors and occupations and in emerging sectors and occupations. To understand what these 'skills' transitions may look like as we learn our way from minimal Just Transitions to transformative Just Transitions, from green resilience to green revolutions and from instrumental business cases for profit maximisation to transformative business models for sustainable value creation, it is useful to explore the implications for particular cases. Case examples provide specific insights into how the principles, processes and practices of these transitions may take place and the kinds of occupations and skills that will be needed (Ward, 2018).

Framing green skills research in a transforming space

Skills development plays a crucial role in seizing opportunities and unlocking the potential of green jobs, with careful attention to the implications if a transformative approach is taken. Planning for skills to support transition to a green economy in developing countries is an important enabling factor. Some generally used definitions of green skills include:

- Green skills are the abilities, values and attitudes people need to build and support a sustainable and resource-efficient society (Cedefop, 2012).
- Green skills are the technical skills, knowledge, values and attitudes needed in the workforce to develop and support sustainable social, economic and environmental outcomes in business, industry and the community (McDonald, Condon, & Riordan, 2009) (Australian Green Skills Agreement, 2009).
- Green skills can be defined as the skills needed by the workforce, in all sectors and at all levels, in order to help the adaptation of the products, services and processes to the changes due to climate change and to environmental requirements and regulations (OECD & European Centre for the Development of Vocational Training, 2014).

A further point absent from many international discourses is that green skills are for the common good.

Drawing from the above examples, within South Africa we needed a typology relevant to the South African skills system that resonated with the South African occupations framework; we have found it useful to categorise the greening of work as shown in Figure 12.3.

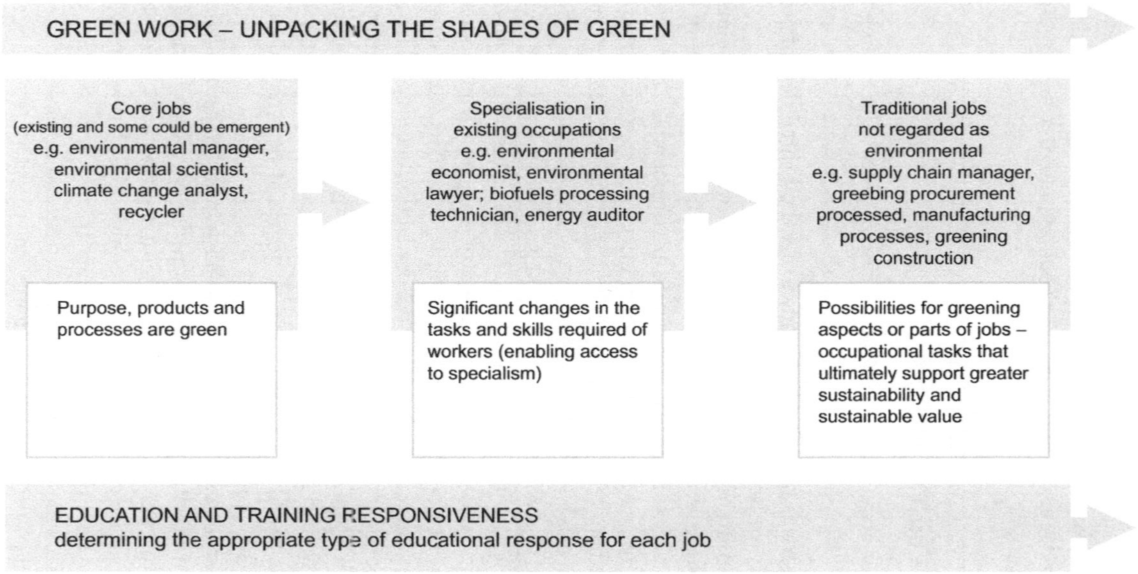

Figure 12.3 Framing the greening of work for the South African occupational framework

This enabled us to characterise the types of changes that were occurring and to make judgements about the educational intervention required. The framing resonated with occupational systems so emerging specialisms and new knowledge could easily be captured.

The case example in Box 12.1 is drawn from a Green Skills in Petroleum study conducted by the author and colleagues (Ward, Jenkin, Rosenberg, Ramsarup, 2016).

In order to formulate differentiated recommendations to diverse sectors studied, we used a tiered approach that could represent a staggered approach to the green transition.

Box 12.1 Case example

Planning for occupational change: A case of petroleum – three tiers of green skills

The Green Skills in Petroleum study focused on three tiers of green skills development within the petroleum sector in South Africa.

Tier 1 refers to those jobs that relate to improving the environmental performance of the sector. These include the continued improvement of energy efficiency practices in oil refineries; waste management; the reduction of fugitive emissions and the production of cleaner fuels. It is likely that over the short to medium term most of the greening of jobs in the petroleum sector will occur at this level as existing refineries are maintained; new refining capacity related to clean fuels is installed and gas-to-liquid capacity is enhanced.

Tier 2 refers to the creation of new jobs that are closely aligned with the existing structure and operations of the petroleum industry. The Department of Energy in South Africa, for example, highlights the creation of jobs through increased biofuel production activities. This approach has a number of advantages related to job creation in South Africa in terms of the cultivation of appropriate crops and the initial extraction of fuel feedstocks from the biomass produced. Although it is unlikely that new occupations would be created, new skills would be required in terms of working with biofuel feedstocks.

Tier 3 refers to a significant shift towards low or zero carbon based energy sources such as wind, solar and wave and the associated increase of alternative energy storage mechanisms. This transition will have significant impacts on some occupations and will require skills development planning and implementation to ensure that this transition is timeous and just. There is a growing trend amongst cities to set target dates for the removal of fossil fuel powered vehicles and a recent report (Arbib & Seba, 2017) suggests that many countries (South Africa is likely to be a laggard but will be affected by global trends) will see a dramatic increase in autonomous electric vehicles. This will, according to the report, mean that conventional energy and transportation industries will suffer substantial job loss. Policies will need to mitigate these adverse effects.

Occupations, occupational systems and green skills analysis

Transitioning to a greener society has focused primarily on shifting technologies and policy. Attention to how this will change the nature of work is rarer and an analytical construct that provides insights into how the nature of work itself is/will change is rarer still. In our research, we have found that occupations as an organising construct for work can serve as a useful analytical construct to analyse the greening of work (cf. Chapter 5). This is not to essentialise occupations discourse in green skills research – as shown in this book there is much more to green skills research than a focus on occupations only. It is also important to note that occupations are inherently transient and not permanent constructs; they exist within wider societies and environments and suffer from internal tensions and contradictions. They are also potentially in tension with other economic and social interests. As shown in the studies that focus on occupations in this book (cf. Chapters 3, 4 and 5), all occupations have both a hierarchical and a network dimension, which enables the study of educational and occupational progression as well as families of occupations and networks. Occupations, as historically forged social constructs, effectively operate as an organising regime which sets out rules of practice and governs qualifications, learning and practice in the field. An intensive look at occupations across a range of sectors and fields can enable a focus on the regime level and surface significant lock-ins that are setting the rules of practice (e.g. the dominance of fossil capital in defining 'relevant' occupations, e.g. in the mining sector) and thus to look at spaces and options for transformative changes. Hence there is need for multi-levelled and systemic analysis around occupations as units of analysis.

Occupations and the defining of occupations are a critical element in trying to understand green work and learning. As can be seen from many sectors other than the environmental sector (at present), underpinning work and learning is coherent occupational framing of sector based occupations and skills provisioning. For the environmental sector, both of these are dependent on a co-ordinated approach to representing occupations in the environmental sector in the official systems of occupational framing. The Organising Framework for Occupations (OFOs) is a "coded occupational classification system" (Department of Higher Education & Training [DHET], 2015). Its emergence stemmed from the challenge of trying to find a common language and vocabulary to report, classify and describe occupations. Without a common language, it is possible to get data on supply and demand from the labour market but it is not possible to interpret and analyse the data to understand the patterns and skills development implications. The OFO is thus a key tool for identifying, reporting and monitoring skills demand and supply in the South African labour market, and its categories are used, for example, in the Labour Force Survey and in the collection of national census data. The OFO lays the systemic framework for occupational progression: occupations are grouped into occupational families and there are feeder occupations for particular occupations. The OFO thus lays a foundation for organisations and sectors to map and outline occupational progression, which is a critical element underpinning green work and learning, especially in a sector with unclear pathways as depicted in Chapters 6 and 7.

In a system that is trying to respond to education and training needs from labour market perspectives, the 'language of occupations' has become the dominant code of meaning within the Occupational Qualifications Framework. It holds power in that resources, qualifications development and skills planning are centred on the language of occupations. Identifying absences such as those identified above start to intensify concern around systemic articulation within an occupationally driven system of skills development, and it is critical they are addressed if meaningful learning pathways as envisaged by the DHET (2012) are to emerge for this new area of economic development and social-ecological practice in South Africa. As indicated in the DEA 2010 study, environmental learning pathways are not only necessary for economic development within the green economy framework, but are also essential to the public good, as they are required to manage South Africa's natural resources and biodiversity sustainably for current and future generations as outlined in the South African Constitution (Republic of South Africa, 1996). Many green skills studies require an understanding of how occupations are framed, how they are represented and how it may be possible to plot/explore or understand occupational hierarchy and possible progression. This is central to understanding occupational progression.

The studies reported across this book suggest that a key starting point for addressing improved learning pathways for green skills or environmental jobs in the sector involves:

- Clarifying environmental occupations by getting accurate reports from employers on the demand and supply of jobs
- Consolidating information on environmental occupations that can be mapped to the OFO and used to correct erroneous mappings
- Comprehensive representation of environmental occupations on the OFO based on clear descriptors of differentiated occupations and associated progression pathways (as is the case in sectors such as agriculture and mining)
- Ensuring availability of different education and training options within a more co-ordinated systemic framework for green skills development that addresses the imbalance between existing high skills and the rest of the skills system.

A more coherent co-ordinated approach that is inclusive of the above processes will potentially assist existing environmental practitioners to meet their skills development needs and enable easier pathways for new entrants to the sector.

A critical analysis of the OFO in South Africa has raised some generic issues around representation and framing of environmental occupations which we need to engage with to enable better movement, progression and navigation within and between environmental occupations.

- Due to its 'newness' as well as the lack of a co-ordinated approach, there is very little available knowledge on the profile of environmental occupations and systemic integration of them into broad occupational frameworks. This has resulted in huge contextual variance in occupational titles, occupational tasks, entrance requirements and general sectoral knowledge of an occupation, which has created difficulty in finding a common language or reference framework for discussion on employment trends and statistics in the sector.
- Representations of environmental occupations are scattered and clear scaffolded pathways are not visible across levels of skill.
- Representations of socio-ecological occupations in OFO are poor – many environmentally linked occupations currently represented are scientists requiring a science learning pathway. Occupations linked to community conservation, social ecology, stewardship and rural development are not represented, thus restricting movement for students with a social sciences/ arts pathway.
- Most entry points in terms of represented occupations are at professional levels, mostly requiring postgraduate qualifications. Greater representation of entry level environmental occupations to enable better access into the sector at artisan and technical skill levels needs to be explored. This would have implications for technical and vocational education and training (TVET) pathways to enable better artisan linked education and training opportunities.
- New fields of occupational practices are not well captured in the OFO, thus skills provisioning and training for these will be reactive, ad hoc and uncoordinated.

The notion of 'greening' the economy will require the identification of particular green jobs, occupations and skills.

Assessing and anticipating these jobs, occupations and skills will require research relating to at least four different situations:

- **Core green occupations:** Some existing occupations are already focused directly on green issues often with a high degree of specialisation. Some examples are Environmental Managers (134901) and Air Pollution Analysts (213305). The primary research question for this group is how many are needed given different supply and demand considerations.
- **Changing occupations:** Within some existing occupations, emerging environmental challenges and opportunities are resulting in a requirement for new skill sets that differ significantly from those required for more traditional work in these occupations. For example, the instillation of solar water heaters (642602) overlaps with more traditional plumbing (642601) work but differs in the specifics of its core skills requirements. Research questions need to establish both the supply and demand of existing skills and qualitatively assess and anticipate the new skills requirements.

- **New skills needs across occupations:** These are existing occupations that do not have a focus on environmental issues but require supplementary skills as environmental issues emerge or are acknowledged in a particular area of work. An example may be the incorporation of sustainability considerations (e.g. water or energy use) into the job of a Facilities Manager (143901) (cf. Chapter 4) or sustainable investment considerations for an Investment Manager.
- **Newly emerging occupations:** Here the requirement is for people with new skills in roles that amount to new occupations. This group is extremely difficult to deal with in existing skills planning frameworks since the occupations have not been defined and, in many instances, even the skills are still emerging in response to new demands. The research for this group is largely qualitative as anticipatory approaches to skills planning seek to clarify what these occupations will look like in the near or medium term. Examples include carbon accounting, chief sustainability officers and alien species control officers.

Using the actors identified within the value chain analysis (cf. Chapter 3), the four categories outlined above and occupational information derived from the OFO, it is possible to develop a map of occupations and occupational families.

The green occupations and green skills analysis is intended to identify the key occupation and skills required in the core and support processes within a value chain. This involves listing all actors and then aggregating them into common categories, in this instance usually occupational groups, and identifying the main environmental issues and skills required to deal with these issues. This process results in a list of occupations and related skills requirements that can be used in different ways by different audiences. It must be stressed that the actor mapping will define the subsequent boundaries of occupational analysis, therefore in producing the actor map it is important to ensure that the scope is commensurate with the project's ability to generate meaningful insights into occupations and skills. This is helped significantly by the use of environmental hotspot analysis to identify priority areas of focus and related actors, occupations and skills.

The first step is to use value chain analysis to identify the main actors and their areas of operation. Building on the value chain analysis and the environmental hotspot analysis, Table 12.1 shows a section of a table that extends over almost four pages (listing approximately 50 occupations) and is linked to the surface coating value chain (cf. Chapter 5).

The next step is to use the value chain and the environmental hotspot analysis (cf. Chapter 4 for an explication) along with interviews and site visits to identify key areas of focus in terms of actors, occupations and skills. These key areas of focus will usually be people who, working individually or as a network, have the potential to make a significant difference in terms of addressing environmental hotspots.

Table 12.1 Surface coating value chain

Value chain process stage	*Process and business activities*	*Main environmental issues (greatest area of impact)*	*Suggested occupations for greening (code)*
	Technology development		
Paint manufacturer	R&D	• Hazardous materials (environmental and health alternatives • Water-based alternatives to raw materials	R&D Manager 122301 Chemical Engineering Technician 311601 Chemical Engineering Technologist 314502/ Chemical Engineer 214501/ Analytical Chemist 211301
	Manufacturing and production		
	Procurement and supply chain	• Green Procurement • Hazardous materials (ingredients)	Procurement Manager 132401 Supply Chain Manager 132401

Source: Jenkin et al., 2016

In the surface painting study (cf. Chapter 5), for example, it became apparent that it was important to focus on three 'occupational networks'. These were: Safety, Health, Environment and Quality; Technical and Laboratory (R&D); and Painters and Contractors. The environmental or greening tasks that these clusters performed (or were increasingly required to perform) were then listed against a broad set of key occupations derived from the OFO.

Key occupations are then examined in detail using the OFO, international equivalents such as the US based ONET and alternative local equivalents such as competency tables developed within particular institutions or industry bodies as reflected in Table 12.2. It is here that the categories of 'dark green', 'light green', 'green tint' and 'new green' are useful for highlighting emerging focus areas that may not yet be included in the OFO. More specifically, this examination of occupations and skills through the various green lenses should lead to a better understanding of whether new skills need to be added to existing occupations that were already sensitive to environmental issues or whether a more fundamental engagement with sectors that were not currently incorporating green skills into particular occupations is required. A small section of the resulting table for the Environment(al) Officer from the surface coating study (Jenkin et al., 2016) is presented in Table 12.3.

Based on this review and the resulting tables, it should be possible to develop a number of conclusions and recommendations regarding key occupations and

Table 12.2 Key occupations and core green tasks

Key occupations in this network	*Core green tasks for this network*
• SHEQ Practitioner • Sustainability Manager • Environment Officer • Environmental Health Officer • Health and Safety Manager or Officer	• Develop or execute strategies to address issues such as energy use, resource conservation, recycling, pollution reduction, waste elimination, transportation, education and building design • Drive lean management throughout the business and raise awareness accordingly • Application of compliance with environment, health and safety legislation and regulation, e.g. ensuring workforce wear protective clothing and masks

skills. In the case of 'dark green' occupations, this is most likely to be focused on the number of professionals required and the additional skills that may be required that are not yet incorporated within the OFO. For 'light green' and 'green tint', recommendations are likely to focus on how particular existing occupations may need to link within networks and what skills would be required to support this. The most challenging recommendations are likely to be around those occupations that are emerging or may be required in the future. Here significant engagement may be required to lobby industry bodies and higher education institutions to name these occupations, ensure they are recognised on the OFO and articulate the skills required within this emerging occupation.

Re-imagining green work

A significant element in surfacing transformative praxis possibilities focuses on strengthening the occupational framework for green work at different skill levels, strengthening information and access opportunities into green work and, finally, a more coherent responsive policy environment that prompts better links between skills and environmental policy. Figure 12.4 also highlights some of the significant elements emergent from the research across this book that present transformative opportunities to enable re-visioning towards a more 'joined up' system for green skills.

Conclusion

Green skills research is shaped by a range of complex drivers or 'underlying mechanisms' – political, economic, social, environmental, social justice and long term ethics. The core knowledge is fuzzy; it exists in multiple pockets and centres of activity and tends to exist in and within transitive contexts. This has methodological implications that demand research methodology with ongoing contextual analysis and review; combined historical analysis and futures thinking;

Table 12.3 Codes, skills and shades of green for Environment Officer

Occupational name: Environment Officer			
OFO major group (code)	*OFO sub-major group (code)*	*OFO Unit group (code)*	*OFO occupation code*
Professionals (2)	Physical, Mathematical and Engineering Science Professionals (21)	No unit group 2133	213302
Segment of the value chain	*Skills band (high/intermediate/low)*		*Shade of green (dark/light/tint)*
Manufacture	High or Intermediate		Dark
Occupation description and purpose	Studies and develops policies and plans for the control of factors which may produce pollution, imbalance or degradation of the environment.		
Greening the occupation recommendation	Field of knowledge	• Understands environmental legislation and standards, e.g. SABS, and application thereof. • Understands the implications of not being compliant. Gets input from legal department regarding updates on compliance (includes national and local by-laws) • Needs to follow and apply environmental management systems (EMS) in line with ISO14001:2004. Environmental Management Systems programme. • Has ability and understanding to interpret environmental initiatives and implement these in the business. • Waste management – map waste arisings and reasons, identify solutions for reduction along the supply chain • Energy management and identification of solutions – where is energy being used, measure and identify solutions – lean manufacture, installation of renewable energy solutions, e.g. solar panels CHP.	

the combination of humanities, social justice and environmental sciences, as well as labour analysis.

Furthermore, the changes in the content and structure of work are shaped by environmental trends that include changes in markets, technologies and workforce demographics. The ability of occupational systems to be responsive to changes is critical. Green skill analysis systems need to engage with occupational systems so that framings of occupations and jobs are coherent and consistent.

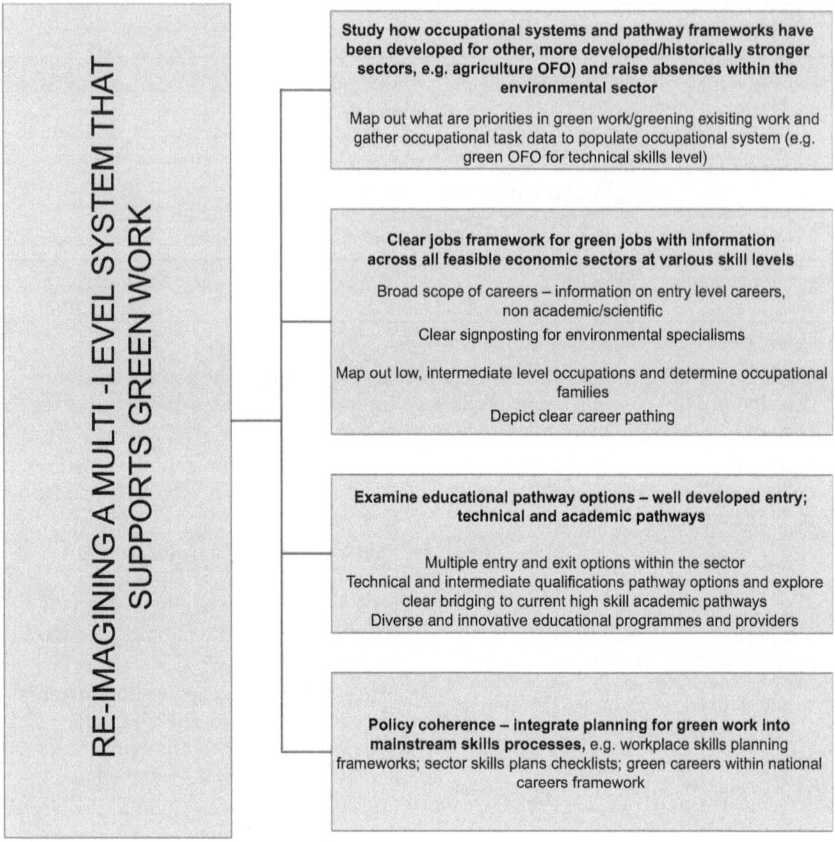

Figure 12.4 Re-imagining systemic support for green work

References

Arbib, J., & Seba, T. (2017). *ReThinkX*. Retrieved from https://static1.squarespace.com/static/585c3439be65942f022bbf9b/t/59f279b3652deaab9520fba6/1509063126843/RethinkX+Report_102517.pdf

Australian Green Skills Agreement. (2009). Council of Australian Governments: Green Skills Agreement: An agreement between the Australian Government and the State and Territory governments. Barton, Australian Capital Territory: Council of Australian Governments.

Cedefop. (2012). Green skills and environmental awareness in vocational education and training. Retrieved from www.cedefop.europa.eu/en/publications-and-resources/publications/5524

Cock, J. (2012). The 'green economy': A just and sustainable development path or a 'wolf in sheep's clothing'? *Global Labour Journal*, 5(1). doi: 10.15173/glj.v5i1.1146

Death, C. (2014). The green economy in South Africa: Global discourses and local politics. *Politikon*, *41*(1), 1–22. doi: 10.1080/02589346.2014.885668

Death, C. (2015). Four discourses of the green economy in the global South. *Third World Quarterly*, 36(12), 2207–2224. doi:10.1080/01436597.2015.1068110

Department of Environmental Affairs (DEA). (2010). *National Greening 2010 Framework*. Retrieved from www.environment.gov.za/sites/default/files/docs/nationalgreening_2010framework.pdf

Department of Higher Education & Training (DHET). (2015). *Complete OFO Version 2015*. Pretoria: DHET.

Economic Development Department (EDD). (2011). *Green Economy Accord*. Pretoria: EDD.

Ferguson, P. (2015). The green economy agenda: Business as usual or transformational discourse? *Environmental Politics*, 24(1), 17–37. doi:10.1080/09644016.2014.919748

Georgeson, L., Maslin, M., & Poessinouw, M. (2017). The global green economy: A review of concepts, definitions, measurement methodologies and their interactions. *Geo: Geography and Environment*, 4(1), e00036. doi:10.1002/geo2.36

Gibbs, D., & O'Neill, K. (2014). The green economy, sustainability transitions and transition regions: A case study of Boston. *Geografiska Annaler: Series B, Human Geography*, 96(3), 201–216. doi:10.1111/geob.12046

Jenkin, N., Ramsarup, P., Rosenberg, E., & Molebatsi, P. (2016). *Green skills in the South African surface coatings sector: A focus on paint*. Research report for Chemical Industries Education & Training Authority. Retrieved from www.greenskills.co.za/wp-content/uploads/2015/07/FINAL-Green-skills-in-the-paint-sector.pdf

Maia, J., & Giordano, T. (2011). *Green jobs: An estimate of the direct employment potential of greening the South African economy*. Retrieved from www.idc.co.za/projects/Greenjobs.pdf

McDonald, G., Condon, L., & Riordan, M. (2009). *The Australian Green Skills Agreement Policy and industry context, institutional response and green skills delivery*. TAFE Directors Australia.

Organisation for Economic Co-operation and Development (OECD), & European Centre for the Development of Vocational Training. (2014). *Greener skills and jobs*. Retrieved from https://doi.org/10.1787/9789264208704-en

Republic of South Africa. (1996). Constitution of the Republic of South Africa. Pretoria: Government Printers.

Unmüßig, B., Sachs, W., Fatheuer, T., & Unmüßig, B. (2012). *Critique of the green economy toward social and environmental equity*. Berlin: Heinrich-Böll-Stiftung.

Ward, M. (2018). *Developing competencies for a just transition of the South African banking sector: Understanding just transition and its implications for the banking sector to inform a consideration of occupations and skills needs*. Johannesburg: Centre for Researching Education & Labour (REAL), University of Witwatersrand.

Ward, M., Jenkin, N., Rosenberg, E., & Ramsarup, P. (2016). *Green skills in petroleum*. Johannesburg: Centre for Researching Education & Labour (REAL), University of Witwatersrand.

13 Synthesis and elaboration of critical realist methodology for green skills research

Eureta Rosenberg

Introduction

The studies shared in this book have all been informed by the philosophy of dialectical critical realism. Critical realism does not prescribe or promote only one kind of research design (Sayer, 2010). Many different types of studies – destination/tracer study surveys; in-depth case studies; combinations of surveys and case studies; life histories; ethnographies; comparative analyses; interventionist studies; and theoretical-conceptual explorations, to name a few, can all be undertaken, with the design decisions informed by the philosophy of critical realism, as developed by Bhaskar (2010) and elucidated by Norrie (2010) and Price (2016).

Why critical realism? It is our experience, in designing and conducting research, that critical realism usefully underlabours for the social sciences, as was first proposed by Bhaskar (1979) in his book *The Possibility of Naturalism* where he deconstructed the dualisms that permeate modern social sciences.

Underlabouring is the process of clarifying or 'clearing away' ontological and epistemological confusions and uncertainties which often arise when research design is approached from a variety of different perspectives (Bhaskar, 2016). Without such clarification, scholars 'talk past each other' on matters such as the validity of different approaches to social science. This is because there are a variety of different (epistemological) standpoints on what counts as science, a terrain so contested that it reminds one of 'tribal warfare' (Pawson, 2006; Janse van Rensburg, 1996). While this chapter is not prescriptive as to the types of research design most suitable for green skills planning, it does illustrate the critical dialectic realist considerations that 'did the clearing' for the studies in this book. Ontological, epistemological *and axiological* ground clearing supports a transformative research intent (Bhaskar, 1993; Norrie, 2010) and a commitment to credible science.

To structure the chapter, we adapt Maxwell's four considerations for planning research (Maxwell, 1996) as follows:

- Political-institutional ('big picture') context
- Theory and conceptual considerations
- Methodological considerations
- Personal-professional (in context).

The political-institutional context of green skills research

Chapters 1 and 2 sketch a broad political-institutional context for green skills research, consisting of ecological, economic and social justice issues, as well as the policy responses at national and international levels. The need to develop and plan for 'green' skills, and to inform such planning and development processes with transformative research, stems from growing reflexivity about a 'polycrisis' that includes unsatisfactory educational outcomes; poverty, reduced and job-less growth, unemployment and inequality attributed to failures of capitalism (Moore, 2015); all of which are exacerbated by climate disasters, dwindling and degraded natural resources and the escalating loss of species. In Chapter 1 we noted the notion of a 'safe operating space' for humanity (Raworth, 2017) which highlights that while we are overshooting planetary boundaries in the biosphere, on land, in freshwater systems and the ocean, through the levels of pollution, extraction and consumption characterising the modern world, the basic needs of the majority of the planet's citizens remain unmet. Societies, social movements and governments are starting to reflexively recognise (Beck, 1992) and respond to these problems. Many of the studies profiled in this book highlight possibilities for change: circular economies in agriculture (cf. Chapter 4) and chemicals (cf. Chapter 5), rehabilitation in mining (cf. Chapter 3), clean energy (cf. Chapter 9) and rainwater harvesting (cf. Chapter 11) to name a few.

Changing the work we do is one of the most significant responses or leverage points (Meadows, 1999) towards system wide transformations to reduce planetary degradation, (re)generate livelihood opportunities and increase human well-being (Ramsarup, Rosenberg, Lotz-Sisitka & Jenkin, 2019). 'Green work' requires skills. Exactly what those skills may be is a matter for research. This research intends to help change the status quo towards well-being for all and ecological sustainability. For this reason, we frame green skills research as transformative or even more strongly, as emancipatory, in the sense of a quest to free society from the assumptions and practices that prevent us – globally, nationally and locally – from operating in an ecologically safe and socially just space (Raworth, 2017).

The resultant *praxiological* (ethical-reflexive-practical) purpose of green skills research has implications for how research design in general is approached, even as more specific contextual factors further define the focus of specific studies. Examples of how this context and purpose shape research design can be found throughout the book, and are pulled together in this chapter.

Given the raison d'être of green skills research sketched above, its relevance in an applied context is a key consideration. What can researchers do to enhance the relevance of their research? One obvious strategy is to formulate the research focus and scope *with* partners who are most likely to be expected to use it. For example, in a study on skills for greening public procurement (Ward, Jenkin, Rosenberg, & Ramsarup, 2016), the researchers formulated the research questions and focus with the Research Manager in the Public Service Sector Education and Training Authority (PSETA) and as a result, the

PSETA retained an interest in the outcomes of the study and continued to open doors for the researchers to present their findings, or share them with training providers, including the University of South Africa and the National School of Governance.

Researchers in the National Environmental Skills Forum found that research was most likely to be experienced as relevant, if it was shared at a time and in a format when partners were looking for strategic and/or policy guidelines. This occurred, for example, when South Africa's Department of Higher Education and Training called for inputs into the National Skills Development Strategy (NSDS), or when the Sector Education and Training Authorities (SETAs) subsequently responded to the NSDS requirement to identify the skills implications of the green economy in their respective sectors. On such occasions, research relevance was enhanced, and available research was translated into policy briefs and guidelines (DEA, 2010; Ramsarup et al, 2017).

Partners' engagement with research findings was stronger in co-engaged enquiry processes, i.e. where researchers and other partners approached the research problem together as a matter of mutual interest. The formative interventionist co-enquiry into Amanzi [Water] for Food in South Africa's rural landscapes, involving the university, the agricultural college, farmers and others (see Lotz-Sisitka & Pesanayi in Chapter 11), is a pertinent example. This generative study resulted in curriculum changes in the form, for example, of new demonstration sites on the college campus and new training programmes. The T-Learning programme discussed in Chapter 10 (Green skills supply: Research from providers' perspective(s)) is a further example of co-engaged research with high levels of relevance for a range of partners.

Theory and conceptual considerations

Transformative green skills research design is best served by an explicit attention to theory. The political-institutional context outlined above may suggest that green skills should focus on producing practical recommendations, and that an engagement with theory would be superfluous. However, it can be argued (with Norrie, 2010, and others) that if the theory guiding a study is not explicit – and critically engaged – then it remains hidden, because there is usually a theory or rationale driving methodological (and analytical as well as political) decisions. If the theory is unarticulated, it is difficult for the researcher and others to critically engage with it (hence reducing inter-subjective objectivity and the trustworthiness of findings (see below). If it remains unexamined, the mainstream theory of the day is likely to take precedence, even if it does not serve the need for transformation (Norrie, 2010).

How do we become explicit about theory?

It is useful to distinguish between two types of theory. The first is meta-theory, such as the philosophies about ontology and epistemology that shape research design decisions at the most fundamental level, for example, what counts as valid knowledge. The second is substantive or domain theory, which guides the choice of concepts and constructs and the way they are approached (Cruickshank, 2003). Both meta-theory and substantive or domain theory help researchers to use a variety of constructs and concepts in a coherent rather than disjointed way, and to make congruent research design and analytical decisions.

Take, for example, the constructs of 'development', 'skill' and 'demand'. These can all be interpreted from an (economistic) human capital development perspective, or from a (critical emancipatory) social justice perspective. From a human capital perspective, the metaphor of a 'skills pipeline' is often used, implying that learning and moving into work and the economy are linear processes. Green skills studies of environmental scientists moving into specialisms and 'boundaryless careers' (cf. Chapter 6) and wildlife vets and municipal workers (cf. Chapter 7) moving from studies into work show that pathways are not linear and that multiple contextual mechanisms, some with deep historical roots, enable or constrain the pathways and choices of individuals. This suggests that journeys into and through work are anything but smooth, straight pipelines insulated from outside influence. This is confirmed as a general contemporary phenomenon of work in the literature cited in these chapters, and Chapters 6 and 7 demonstrate the need for and availability of theory that is able to capture the dynamic interactions between individual and system that shape learning and career pathways. The work across this book surfaces the need to move beyond 'pipeline' metaphors; a more apt 'supply' model would be to view learning and career pathways as embedded in multiple overlapping and transitioning systems (both industry specific and broader social, economic and environmental changes – cf. Figure 1.4 in Chapter 1 and Figure 14.1 in Chapter 14). The latter perspective more strongly supports a critical emancipatory perspective, as it highlights the many factors that need to change to allow for more transformative skills development for green work, and to 'facilitate and 'ease' access and progress for especially those with lower and mid level skills, as studies show that they have the harder time of it (cf. Chapters 6, 7 and 8).

In Chapters 2 and 12, Mohamed and Ramsarup explain that the concept of the 'green economy' can also be (and indeed is) differently interpreted. For example, there are technical-industrial framings of the green economy, and political-sociological framings. The choice matters. Governments will make different policy and investment decisions, based on the chosen green economy theory or model. In the case of research, a researcher working with emancipatory intent, using theories based on keeping the status quo intact, will not only lay themselves open to critique about inconsistencies, but may also fail to achieve an emancipatory or generative intent, on completion of the research.

The choice of theoretical framing is seldom simply a matter of researcher preference. The analysis by Raworth (2017) shows that mainstream economic theory tends to exclude households and the commons from its ambit, and to position the market as best off with minimal governance (liberal free market economics). From a social-ecological justice perspective, governance by the state, the resources of the commons (e.g. water) and households all contribute significantly to the success of markets. Therefore, to exclude them from economic theory is not merely a matter of preference; it is an error, which renders the theory less reality-congruent than it could be, and likely to result in various failures if applied as is.

In keeping with this trajectory, Ramsarup (cf. Chapter 12) argued, for example, for the inclusion of a focus on occupations in green skills research, without which the findings are less likely to be taken up into systems that are driven by this 'unit of analysis' and are therefore less likely to effect significant systems change. She further demonstrated, in the same chapter, how basic concepts like 'green jobs' can be interpreted in different ways and how the transformative intent shapes particular framings that will in turn assist researchers to work in a transformative manner. Despite its importance, a focus on work and skills is often neglected in studies with an explicit intent to address unemployment and climate threats to economic growth. An example is the Climate Smart Agriculture studies in the Western Cape, on which Cobban and Visser (2017) drew for their green skills study (discussed in Chapter 4). These skills researchers also found that when they changed the framing of their study from Climate Smart Agriculture to a broader social justice angle, which included climate considerations as well as employment, it was easier to see strategic leverage points to intervene in the system.

The choice of meta-theory is also important. In the works gathered in this book, the meta-theory of dialectical critical realism helped the researchers to reorient or adapt substantive theory to expand its emancipatory power, through imminent critique or interrogation of key concepts that are routinely used uncritically in the field of skills development. For example, the studies in mining, agriculture and chemicals (cf. Chapters 3, 4 and 5) interrogated the manner in which industry value chains are routinely presented as linear processes without governance (regulation and enforcement) or supportive organisational processes, and without attention to the 'end-of-life' phase of commodities. In Chapter 8 Lotz-Sisitka explored the notion of a skills ecosystem and demonstrated that an expansion of a 'market based' skills ecosystems to a 'social ecosystem for skills' (cf. Hodgson & Spours, 2018) gives the existing theory more reality-congruence, as well as greater explanatory and generative/transformative power. This example of a theoretical framing (cf. Chapter 11) demonstrates the value of theory for synthesis across multiple studies. This point is well made by Sayer (2010) who drew on Bhaskar's stratified ontology developed in basic critical realism, to highlight the value and indeed necessity for multiple types of studies (concrete, abstract, empirical and theoretical) in order to adequately access the diverse layers of social realities.

Methodological considerations

A laminated approach to green skills studies

A laminated approach to enquiry sees researchers addressing the research interest by investigating several domains of reality that could be viewed as levels, *in relation to each other*. Examples of different domains of reality are macro, meso and micro levels of a skills system (cf. Figure 3.1 in Chapter 3), or the embedded systems of the individual, organisational, sectoral and broader community (cf. Figure 1.4 in Chapter 1). While many studies look at issues at both a macro level and a micro level, it is the relationality of these, or the resolution of a commonly formed dualism, that is significant here (cf. Chapters 6 and 14).

The laminated approach derives from the realist underlabourer. Realism is the philosophy that an intransient reality exists (the Real), beyond (but including) the more transitive realities of what humans may think exists, beyond (but including) what we happen to observe (the empirical) and even beyond (but including) what actually transpires (the actual).

Critical realism asserts that this reality is stratified such that some layers are 'deeper' than others, that is, they relate to other layers in a causal manner, giving rise to the 'emergence' of new causal mechanisms and outcomes. For example, a region with good water resources will have more development options than a region with low rainfall, so climate is in a causal relationship to development. A particular kind of development would give rise to certain (emergent) employment patterns, for example, the introduction of automation in the mining industry would cause a pattern of growth in technical jobs as opposed to a growth in manual labour jobs.

This example shows that some causal mechanisms are givens of a biophysical nature while others are socio-technological in nature; both are described in Bhaskar's realist philosophy as the reality principle ('natural necessity') (Norrie, 2010, summarised this well). For example, the introduction of automation in mining is a reality, but it is a reality that can change more readily than, say, the climate of a region.

While the point about the existence of structural causes is obvious enough, the design of positivist research is premised on an often implicit assumption that only that which can be directly observed (or even, objectively measured) actually exists (known as the epistemological fallacy, associated with monovalent ontology, or a shallow ontology compared to the depth ontology suggested by social and critical realism). A critical realist philosophy gives rise to a number of analytic processes (see below) through which researchers can access those 'hidden from sight' layers of the totality of the Real, informed by the Realist question: What must reality be like, for the observed to appear so?

The earlier example shows that causal powers (climate and technology) operate at different levels; climate is for the most part more fundamental than technology. They also interact with each other, counteracting or reinforcing each other. For example, technology can be introduced to counter some climatic conditions (e.g. irrigation can counter low rainfall), but such technology can in turn also exacerbate other climate related factors (Bhaskar, 2010).

The researchers producing the works in this book argued that to understand green skills needs, research has to probe these multiple layers of influences, and how the interactions between them are likely to shape (emergent) skills needs. This is captured in the diagrams (developed by Lotz-Sisitka, Ramsarup & Rosenberg) of nested, overlapping systems in transition (cf. Figure 1.4 in

Chapter 1), of iterative enquiry across macro, meso and micro levels of the skills system (cf. Figure 3.1 in Chapter 3) and of work and learning transitions (cf. Figure 14.1 in Chapter 14), all of which draw on Bhaskar's seven scalar laminated system (cf. Figure 14.1 in Chapter 14).

The value of this approach is inter alia in its explanatory power: it helps scientists and decision makers avoid relativism, as researchers are able to identify generative mechanisms that allow for a deeper understanding of the interconnected nature of the world, which includes both transitive and intransitive dimensions (Bhaskar, 1993).

Dialectic and absence

Dialectic critical realism, developed by Bhaskar (1993) out of his earlier basic critical realist project that vindicated ontology in the natural and social sciences, and accessibly interpreted by Norrie (2010, see also Bhaskar, 2010; 2016), is particularly valuable for its transformative power and, therefore, for generative research. The form of transformation we have in mind here can be described as emancipatory: it frees us from problematic, as-yet-unexamined assumptions about, for example, development and environment. Bhaskar (1993) poetically refers to 'the pulse of freedom', to be found in the realist (as opposed to Hegelian) *dialectic*. At the heart of the critical realist dialectic are the notions of totality and absence, and Bhaskar describes learning or engaging with 'what is not yet there' as the 'kernel' of his emancipatory dialectic (Hartwig, 2007). In this view of reality, there is a totality, which is open and consists of partial or sub-totalities, which includes absences or negatives. A positivist philosophy focuses on what is, and ignores what is not; this, Norrie (2010) argued, along with Bhaskar, can be traced to the Greek roots of philosophy, which had an interest to serve the status quo: a society in which the fall of the aristocracy or the freedom of slaves (absences or negatives) were to be avoided as topics for scholarly consideration. For purposes of transformation, in the form of emancipation from an undesirable status quo, there is a need to focus on the negative (absence), including negating the current negatives towards new positives. Absence and absenting (including absenting the absences which keep absences or 'ills' like environmental degradation or economic inequality in place) includes that which is not present here, but may be present elsewhere. Various 'green skills' related studies (cf. Lotz-Sisitka, 2016, Mukute et al., 2016; Ramsarup, 2016; and Pesanayi, 2019) provide further elaboration and examples of processes of 'absenting absences, constraints and ills' as transformative praxis.

We have thus found 'absence' to be a vitally important research construct in green skills research. For example, if a government entity failing to adequately deliver on its difficult mandate to *both* promote mining *and* protect the environment suggests it has no skills needs, there is value in focusing on the absences: the failures in delivery, the absence of an aligned mandate, the absence of recognition of the contradiction and the absence of agreement among key informants.

In another example, structured learning and occupational pathways are absent for environmental work in South Africa, but they do appear in other countries and in other (more established) sectors. Ramsarup (2017) therefore studied the presence of learning and occupational pathways in Agriculture, in order to better understand their absence in the environmental sector (the focus of Chapter 6).

Other examples of the value of focusing on 'absences' and 'absenting' in green skills studies are:

- The expansion of value chains to include governance and regulation (Jenkin, Ramsarup, Rosenberg, & Molebatsi, 2016), as well as the potential (not yet there) of circular economies (e.g. collecting and re-using post-consumer paint, or creating jobs post-mine closure through rehabilitation); these are the case examples shared in Chapters 3, 4 and 5
- Adding the dimension of History to contextual driver analyses which typically would only focus on current Policy, Economic, Social, Technological, Legal and Environmental factors (cf. Chapters 3, 4 and 5)
- Considering the absences in vocational education and training curricula and learning approaches (cf. Chapters 10 and 11)
- Focusing on systems not as static, but as in movement – transitioning (cf. Chapters 6 and 7).

Emergence and a depth approach to complex systems

The relationship between critical realism and systems theory is a matter of scholarly exploration (see, for example, Mingers, 2010) that will not be explored here. Most studies in this book have been approached from a systems perspective, giving explicit attention to relationships between subject and context, multiple interacting dimensions of subjects and contexts and, significantly, to depth ontology, relationality and the possibility of emergence. This approach recognises green skills praxis as occurring in open systems that are also complex, so that they are difficult to bound and outcomes are not easy to predict on linear pathways. Outcomes are, however, not random; they emerge from the stratified layers of reality discussed earlier, and from structure-culture-agency relationships in the social world, as these intersect with and emerge from materiality and history (Archer, 2000). That is, the critical realist view of systems has a depth ontology. As a result it gives an unusual amount of attention to hidden structures and mechanisms including history and power relations, with their presence in the present, as in the mining and agricultural studies (cf. Chapters 3 and 4) which reflect socio-economic lock-ins with deep historical roots, still affecting skills needs today, and the studies of learning and career pathways, where both the possibilities and the absences (of say, articulation pathways for middle level environmental skills, or the presence of adequate science marks in school) can be traced to historical factors, and a complex combination of generative mechanisms that intersect to shape the present (cf. Chapters 6 and 7).

Relationality

Another key concept evident in this book is relationality. It is derived from a complex systems approach (relations among objects, and between objects and context) and a critical realist philosophy (relationships between layers of the stratified Real, multiple interacting mechanisms with causal powers). Relationality is evident in the way in which the model in Figure 1.4 (cf. Chapter 1) is applied, and in many studies in this volume. One example is the discussion of the competency-based learning needs assessment in Chapter 9. Reflecting on this study, to determine the learning needs of green economy 'champions', Lotz-Sisitka (2018) pointed to an individualistic fallacy in the learning needs assessment discourse, which tends to assume that people apply their skills and knowledge as isolated individuals outside of context. Instead, she argued, competencies to respond to complex sustainability challenges and opportunities are distributed among groups of people, and never reside only in an individual. Furthermore, they are brought forth within these collectives in relation to real world challenges, that is, situated contexts. To call individual sustainability practitioners 'champions' or 'pioneers' recognises the importance of agency, but also obscures the collective nature of that agency, which is of course important to understand more deeply, if we are to design interventions that build agency. Vandenberghe (2014, cited by Lotz-Sisitka, 2018, p .67) provides an in-depth analysis of spokespersons in social movements in which, he claims, "the network is collapsed into the person who embodies it … an individual comes to personify a collective or unit". This sociological explanation of agency deploys the notion of relationality. With a realist lens, Lotz-Sisitka (ibid.) highlighted the value of Vandenberghe's analysis of the conditions of the possibility of the group and the champion emerging and how this is mediated, with historical and socio-material factors shaping the possibilities for an individual to exercise agency on behalf of the group (for example, institutional cultures, histories, and power relationships). This exemplifies the depth ontological perspective on relationality and systems.

Analysis

As can be seen across the pages of this book, researchers have drawn on the above meta-theoretical concepts to guide analysis, using a combination of critical realist concepts to underlabour for analysis guided by field-based substantive theory. For example, Chapters 6, 7, 8, 10 and 12 show how researchers use field-based substantive theory such as human development theory, curriculum theory, transitioning theory, learning theory, occupations theory and career theory to interpret specific dimensions of the green skills landscape using inductive and abductive modes of inference (see Table 13.1). In addition, researchers also underlabour these inductive and abductive analyses with retrodictive (also sometimes referred to as retroductive) critical realist analysis (see Table 13.1) to probe shaping mechanisms and absences at work at the level of the Real. This allows them to work out why what they were able to understand through substantive field-based

Table 13.1 Modes of inference

	Induction	*Abduction*	*Retrodiction (also referred to as retroduction)*
Fundamental structure/ thought operations	Draw universally valid conclusions from a number of instances. See similarities in a number of observations (identifications of patterns in data)	Interpret and recontextualise individual phenomena within a conceptual framework (theory) or set of ideas. Understand something in a new conceptual framework	From a description and analysis of concrete phenomena, reconstruct the basic conditions for these phenomena to be what they are, by way of thought operations and counterfactual thinking
Central issue	What is the element common for a number of observed entities?	What meaning is given to something interpreted within a particular conceptual framework?	What qualities must exist for something to be possible?
Strength	Provides guidance in connection with empirical generalisations	Provides guidance for interpretative processes by which to ascribe meaning to events in relation to a larger context	Provides knowledge of transfactual conditions, structures and mechanisms that cannot always be observed in the domain of the empirical
Limitation	Inductive inference can never be either analytically or empirically certain (internal limitations of induction)	There are no fixed criteria from which it is possible to assess in a definite way the validity of an abductive conclusion	There are no fixed criteria from which it would be possible to assess in a definite way the validity of a retroductive conclusion

Source: Adapted from Danermark et al., 2002, pp. 80–81

theoretical analysis, is the way it is. In this way, critical realists can produce more substantive hypotheses that can be further researched in other contexts, or from which they can develop transformative praxis in co-engaged ways.

To guide careful, rigorous and trustworthy analysis, Bhaskar (2016) recommended a process of research (called RRREIC) in the social sciences that uses *resolution* of the problem into its components (e.g. local government workers' experience of sustainable development) and involves *redescribing* it (e.g. into career stories or narratives), and *retrodicting* patterns observed (thought experiments to consider possible antecedent causes that, if they were real, would account for the phenomenon). As it is possible to propose a huge number of possible causal mechanisms influencing a phenomenon in an open system, it is necessary to *eliminate* some of these for a more rigorous account, which helps the researcher to *identify* the most plausible account of the causally efficacious generative mechanism(s) or structure at work. This helps the researcher to *correct* earlier accounts and to propose that the mechanisms identified are plausibly causally efficacious in the problem context. For example, in Chapter 6, Mohanoe through a retrodictive (or retroductive) analysis conducted in this way, proposes that injustices of the past education system, power relations, gender relations and political neglect are causative of the lack of learning pathways for workers involved in green skills in the municipality that she studied. This may be generalisable (at the level of the mechanism, cf. Danermark et al., 2002) to other municipal contexts in South Africa, but her explanations would have to be tested in these contexts, as knowledge is always fallible in open systems.

The process above also helps researchers to identify absences (cf. above), which takes critical realist analysis into a generative dynamic, where absence can point out pathways for possible transformative praxis, to be realised via co-engaged transformative learning research, where learning theories such as those proposed by Engeström and Sannino (2010) can aid with emergent process analysis (cf. Chapter 11). Bhaskar (2016) referred to this as a *Diagnosis*, *Explanation*, *Action* (DEA) analytical approach suitable for dialectical critical realist progressive change research processes, which build on the RRREIC approach.

The personal-professional in green skills research

Given the strong emphasis in this book on the *intentionality* of green skills research, this final section will focus on the credibility of research in relation to researchers' personal-professional commitments. Do researchers' values influence methodological decisions? Should researchers be neutral?

Historically, Western science has been characterised by an actualist ontological position (based on Hume), with the epistemological implication that there is a truth out there – in this case, about green skills – to discover, and that a researcher needs to be value-free and objective in order to make such 'out there' discoveries, as bias would blind one to the truth. The associated positivist research framing has been critiqued for pretending that only that which can be empirically observed exists (epistemological fallacy, ontological monovalence) (Norrie, 2010).

An alternative research tradition, also characterising Western thought, posits that all of social reality is constructed (for example, the notion of green skills is a construct to understand rather than a reality to discover) and that therefore we can only study the meanings made of these constructs. The related approaches (from interpretivism/hermeneutics and constructivism to post-structuralism; often based on ontological Idealism sensu Hegel) have also been critiqued: they easily lead to a relativist position from which no clear decisions can be taken.

Relatedly, it ignores that some intransiences do apply, in both bio-physical and social realms. For example, it is undeniable that fertile soils are washing into the ocean, that species are going extinct, that some economies do not grow and that a majority of people lack secure livelihoods.

As noted, critical realism underlabours for the social sciences by clearing away the ontological debris in the aftermath of convincing critiques of positivism and post-structuralism (Vandenberge provides a good review of these). New ground is opened up (critical realism rather than actualism or idealism), in which one can recognise that while social dimensions such as skills and work are qualitatively different from the natural world, they also share features – often hidden from empirical view – such as a structure, causality and emergence. A social realist framing (Archer, 2000; Bhaskar, 2010) positions researchers as actors within the contexts they are researching and the very act of researching engages and changes the context (to a greater or lesser extent, and perhaps more rapidly so in the social realm than the bio-physical realm). To deny the values of the green skills researcher (and the other research participants, such as the educational providers, government officials or corporate representatives being engaged) is therefore to simply privilege hegemonic values that are, by the very fact that they are hegemonic, almost invisible – we are born into them (Archer, 2000). Standpoint theorists like Harding (2005) argued that researchers should declare 'where they come from' in relation to the research questions, and what their own interests or commitments in this regard may be.

In most of the studies in this book, the researchers designed and conducted their enquiries from within an explicitly articulated value position: committed to education and human development, environmental protection and just social transitions towards a sustainable future for all life on Earth.

Rather than exemplifying an undesirable departure from an ideally neutral-objective researcher, such a value position might be necessary, if research is going to make a contribution to the issues emerging in the Anthropocene. Price and Lotz-Sisitka (2016, p. 1) agreed with Spivak (1999, p. 46) that addressing the social-ecological concerns of our times would seem to require intentional human agency of all 'planetary subjects', "to achieve transformation towards sustainability in ways that ensure the flourishing of all current and future generations". They use Bhaskar's (2008, p. 9) description of agency as "'*radically transformed transformative praxis*, oriented to rationally groundable projects'". Price (2016) reflects that Bhaskar (2010), arguing from the standpoint of concrete utopianism, suggests that a key role for intellectuals and educators is to be involved in the envisaging of alternative possible futures for humanity.

What measures do researchers put in place to ensure that the findings from their studies are new (beyond what was known before by them and others), reality congruent and generative of useful insights and recommendations? Lather (2006) and others have provided the field with a number of alternatives to a positivist framing of validity, which assumes that only neutrality can produce credible, trustworthy research findings. One such alternative is what Harding (2005) called 'strong objectivity', which can be further deepened and complemented by the carefully rigorous process of Bhaskar's critical realist RRREIC methodology outlined above, and researcher reflexivity which Price (2016) described as follows:

> A reflexive agent will therefore try to avoid theory/practice inconsistency; that is, they will try to 'walk their talk' which involves being able to engage in processes that help one to see contradictions, try to explain them, and resolve them in one's own life and the world around one.
>
> (p. 36)

How does research come to valid conclusions in the event of differing opinions? One common strategy is triangulation, that is, deploying multiple data sources, multiple research methods and/or multiple research perspectives (coming at the focus of enquiry from different angles, so to speak). An example from the study on mining (cf. Chapter 3) is illustrative of an alternative (realist) approach. While one key informant (a government official) reported that there were no skills gaps in his department, other informants did see skills gaps in the same department. Conventionally, the researcher would resolve this contradiction through triangulation (using multiple data sources) perhaps arguing that if the majority of informants say there *is* a skills gap, then there is one. However, with the realist laminated methodology deployed in this study, the contradictory finding at the micro level could be explained by the finding of an emergent contradiction among the meso level drivers. Through this contradiction emerged conditions under which graduates approving mining applications were '*on paper*' adequately skilled, whereas '*in practice*' they experienced great (micro level) difficulties in the execution of tasks that were being complicated by these tensions, so that they themselves, and others, reported their skills as inadequate.

Conclusion

This chapter has outlined a number of design considerations and the kinds of decisions researchers would take if they deployed a critical realist approach. Some overarching points to highlight are listed below.

Methodological decisions must be preceded by ontological, epistemological and ethical (axiological) tenets being clarified, while recognising that these may also emerge in the process of research.

A realist, laminated methodology allows for the kind of depth and rigour of analysis outlined above, which seems to be needed to inform skills planning in complex social-ecological systems and settings such as those found in South Africa; these studies are often of a smaller scale given the amount and depth of work that needs to be done.

Nonetheless, strong synthesising framings are also needed to be able to look across these and feed system-level recommendations (cf. Chapter 14).

A willingness to probe more deeply, looking at relationality, absences and the presence of the past in the present, can be particularly generative and demonstrative of the transformative power of explanatory and generative research informed by dialectical critical realism.

Finally, provided careful attention is given to the above considerations, a diversity of research models and methods can be useful for answering green skills questions, including descriptive-analytical studies, interventionist studies, theoretical work and more; this will be picked up in the final chapter (cf. Chapter 14).

References

Archer, M.S. (2000). *Being human: The problem of agency*. New York: Cambridge University Press.

Beck, U. (1992). *Risk society*. London: Sage.

Bhaskar, R. (1979). *The possibility of naturalism*. Brighton: Wheatsheaf.

Bhaskar, R. (1993). *Dialectic: The pulse of freedom*. London: Verso.

Bhaskar, R., Frank, C., Hoyer, G., Naess, P., & Parker, J. (Eds). (2010). *Interdisciplinarity and climate change: Transforming knowledge and practice for our global future*. London and New York: Routledge.

Bhaskar, R. (2016). *Enlightened common sense: The philosophy of critical realism*. London: Routledge.

Cobban, L., & Visser, Z. (2017). *Green skills for climate-smart agriculture: A case study of poultry, winter grains and deciduous fruit value chains in the Western Cape*. Cape Town: ADCI (African Climate Development Initiative), University of Cape Town. Retrieved from www.greenskills.co.za

Cruickshank, J. (2003). *Realism and sociology: Anti-foundationalism, ontology, and social research*. London: Routledge.

Danemark, B., Ekström, M., Jakobsen, L., & Karlsson, J. (2002). *Explaining society: Critical realism in the social sciences*. London: Routledge.

Department of Environmental Affairs (DEA) (2010). *Integrating the environmental driver into sector skills plans: An enabling document for all SETAS*. Retrieved from www.greenskills.co.za/wp-content/uploads/2015/07

Engeström, Y., & Sannino. A. (2010). Studies of expansive learning: Foundations, findings and future challenges. *Educational Research Review*, 5(1), 1–24.

Harding, S. (2005). Rethinking standpoint epistemology: What is "strong objectivity"? In A.E. Cudd & R.O. Andreasen (Eds.), *Feminist theory: A philosophical anthology*, pp. 218–236. Oxford: Blackwell.

Hartwig, M. (2007). Critical realism. In M. Hartwig (Ed.), *Dictionary of critical realism*, pp. 97–98. London: Routledge.

Hodgson, A., & Spours, K. (2018). *A social ecosystem model: Conceptualising and connecting working, living and learning in London's New East*. The East London Vocational Education and Training: Innovation through Partnership Programme (ELVET). ELVET Research Briefing No. 3, March 2018.

Janse van Rensburg, E. (1996). *Environmental education and research in southern Africa: A landscape of shifting priorities*. Unpublished PhD thesis, Rhodes University, Grahamstown, South Africa.

Jenkin, N., Ramsarup, P., Rosenberg, E., & Molebatsi, P. (2016). *Green skills in the South African Surface Coatings sector: A focus on paint*. Research report for Chemical Industries Education & Training Authority. Retrieved from www.greenskills.co.za/wp-content/uploads/2015/07/FINAL-Green-skills-in-the-paint-sector.pdf

Lather, P. (2006). Paradigm proliferation as a good thing to think with: Teaching research in education as a wild profusion. *International Journal of Qualitative Studies in Education*, *19*(1), 35–57. doi: 10.1080/09518390500450144

Lotz-Sisitka, H. (2018). Think piece: Pioneers as relational subjects? Probing relationality as phenomenon shaping collective learning and change agency formation. *Southern African Journal of Environmental Education*, *34*, 61–73. doi: 4314/sajee.v.3:1.5

Maxwell, J.A. (1996). *Qualitative research design: An interactive approach*. London: Sage.

Meadows, D. (1999). *Leverage points: Places to intervene in a system*. Harland, VT: The Sustainability Institute.

Mingers, J. (2010). *The contribution of systemic thought to critical realism*. Working Paper no. 232, November 2010. London: University of Kent.

Moore, J. (2015). *Capitalism in the web of life: Ecology and the accumulation of capital*. London: Verso.

Mukute, M. (2010). *Exploring and expanding learning in sustainable agriculture practices in southern Africa*. Unpublished PhD thesis, Rhodes University, Grahamstown, South Africa.

Mukute, M., Marange, T., Masara, C., Sisitka, H. and Pesanayi, T. (2012). *Future capacity building: Capacity assessment for environmental policy implementation*. Howick: SADC REEP.

Norrie, A. W. (2010). *Dialectic and difference: Dialectical critical realism and the grounds of justice*. New York: Routledge.

Pawson, R. (2006). *Evidence-based policy: A realist perspective*. London: Sage.

Pesanayi, V.T. (2019). *Boundary-crossing expansive learning across agricultural learning systems and networks in southern Africa*. Unpublished PhD thesis, Rhodes University, Grahamstown, South Africa.

Price, L. (2016). Key critical realist concepts for environmental educators. In L. Price & H. Lotz-Sisitka (Eds.), *Critical realism, environmental learning and social-ecological change*, pp. 18–39. London: Routledge.

Price, L., & Lotz-Sisitka, H. (Eds.). (2016). *Critical realism, environmental learning and social-ecological change*. London: Routledge.

Ramsarup, P. (2016). Absenting the absence of parallel learning pathways for intermediate skills: The 'missing middle' in the environmental sector in South Africa. In L. Price & H. Lotz-Sisitka (Eds.), *Critical realism, environmental learning and social-ecological change*, pp. 116–136. London: Routledge.

Ramsarup, P. (2017). *A critical realist dialectical understanding of learning pathways associated with two scarce skill environmental occupations within a transitioning systems frame*. Unpublished PhD thesis, Rhodes University, Grahamstown, South Africa.

Ramsarup, P., Rosenberg, E., Lotz-Sisitka, H., & Jenkin, N. (2019). Chapter 8: Green skills: Transformative niches for greening work. In N. Mohamed (Ed.). *Sustainability transitions in South Africa*. London: Routledge.

Ramsarup, P., Ward, M., Rosenberg, R., Jenkin, N., & Lotz-Sisitka, H. (2017). *Enabling green skills: Pathways to sustainable development: A source book to support planning for green economies*. Grahamstown and Johannesburg: Environmental Learning Research Centre (ELRC), Rhodes University and Researching Education & Labour (REAL), University of Witwatersrand. doi: 10.21504/10962/62694

Raworth, K. (2017). *Doughnut economics: Seven ways to think like a 21st-century economist*. London: Random House Business Books.

Rosenberg, E., Lotz-Sisitka, H.B., & Ramsarup, P. (2018). The green economy learning assessment South Africa: Lessons for higher education, skills and work-based learning.

Higher Education, Skills and Work-Based Learning, 8(3), 243–258. doi: 10.1108/HESWBL-03-2018-0041

Sayer, R.A. (2010). *Method in social science: A realist approach*, 2nd ed. London: Routledge.

Spivak, G. (1999). *Imperatives to re-imagine the planet. Imperative zur Neuerfindung des Planeten*. Vienna: Passagem.

Ward, M., Jenkin, N., Rosenberg, E., & Ramsarup, P. (2016). *Occupationally directed skills development for green public supply chain management*. Grahamstown and Johannesburg: Environmental Learning Research Centre (ELRC), Rhodes University & Researching Education & Labour (REAL), University of Witwatersrand.

14 Green skills research

Implications for systems, policy, work and learning

Heila Lotz-Sisitka and Presha Ramsarup

Introduction

This chapters across the book, together with the transformative research orientation adopted, have surfaced a number of important findings for skills system building, most notably the need to pay attention to cross system transformation, and a move away from fragmentation and atomism in skills research and planning. This is so that changes in one part of the system are connected to and better aligned with changes in other parts of the system, this being at wider, national systemic level (cf. Chapter 12) as well as at regional levels (cf. Chapter 8). We have located this work within the wider framework of just transitions towards sustainability, recognising that the emergence of a national and international skills system that fully reflects and is adequately responsive to the demands of a more just, sustainable society is emerging alongside wider systemic changes (cf. Chapters 1 and 2). For example, the advent of renewable energy technology has created new demands for skills development in the energy sector. This has required design and development of new qualifications, new forms of work and learning and new configurations of work and learning. Similarly, the inherited context of marginalisation and poverty in South Africa created demand for innovative approaches to skills development (cf. Chapter 8). Our early research into green skills emergence pointed to a largely reactive green skills system (Department of Environmental Affairs & Tourism [DEA], 2010) with the International Labour Organisation (ILO) noting that South Africa's green skills system is 'far behind' its environmental policy ambitions (cf. Chapter 2). This problem catalysed a strong need to strengthen green skills research in South Africa.

One of the major problems we identified in early green skills research was that traditional research orientations were not adequate for identifying green skills demand, transitioning, occupational analysis or green skills supply. This is because many of the green jobs are themselves still emerging and because green skills are often 'latent' and not yet articulated in terms of green work and the greening of work (cf. Chapters 3, 4 and 5). Additionally, green skills transitioning between education levels and into work is complicated by the fact that the qualifications and workplace demands are poorly aligned as these are often 'new' or under-developed (cf. Chapters 6, 7 and 8). Green skills occupations are poorly defined in the national system of occupations, as there is as yet under-developed understanding of the nature of green work, greening of work and green jobs (cf. Chapters 3, 4, 5, 8 and 12). From a green skills supply perspective the current education and training system is characterised by a history of exclusion, poor quality and education and training programmes that were designed for apartheid-led extractivist economies, rather than green and decent work trajectories. Learning theories in use are inadequate for the types of transformative learning required for green skills development in a context of wicked problems and emerging sustainability transitions (cf. Chapters 8, 10 and 11). These circumstances have required that we work on developing new methodologies to address the challenges found in the field (cf. Chapter 13). There are many implications arising from this work, which we outline briefly below, noting that these implications also set the direction for ongoing research in this emerging field of educational research.

Differentiating types of green skills research

As noted above, in conducting green skills research over the past 30 years, we have found the need for different types of studies that collectively provide in-depth as well as broader insights into how green skills can be developed and reflected across and within an integrated perspective. We identified three particular types of studies, with different critical issues emerging from each of the types of study that requires further engagement and further research.

Type 1 studies: The first type of study requires *in-depth analysis of specific system elements* such as how competences are developed in new green economy activities (cf. Chapter 9), how green occupations are being described and represented in the organising framework for occupations (cf. Chapter 12), how demand for green skills is being conceptualised and established in different sectors (cf. Chapters 3, 4 and 5), how providers are responding to green skills needs (cf. Chapters 10 and 11), how students and employees are transitioning into new green jobs (cf. Chapter 7) and how existing work is being transformed leading to challenges for transitioning and career development (cf. Chapters 6 and 7) (all of which we have focused on in this book). There are other important system elements that also require in-depth investigation, such as how the National Skills Funding system is responding to green skills demand (Sauls, 2018), how ethics shapes learning programmes and pathways (Olvitt, 2012), how upskilling practices and professional learning is shaped (Bell, 2019), how transformation and transitioning into the environmental sector via higher education learning pathways is being experienced (Dotwana, 2015), how vocational education and training is structured and can be transformed (Pesanayi, 2019) and more. There is also need for evaluation studies that show the potential for green jobs development (Ward et al., 2017) and efficacy and nuance of training programmes (Songqwaru, 2019).

We have learned that in-depth studies of this nature provide particular vantage points on a system of skills development, and, as shown in this book, it is the collective vantage point across the studies that begins to develop a wider view of education system change within a wider just transitioning framework. As such, each of these studies offers useful insight into the structuring of a national system of skills and the dynamics of how green skills are currently being advanced and integrated into the system or not. Overall, the findings of these types of studies show that there is movement towards integration of green skills in the national post-schooling system, but that it is difficult to see and appreciate the cumulative impact as much of the innovation is bottom-up, and shaped by emerging responses to changes in context, rather than systemic, joined up system wide planning and alignment.

Furthermore, this research offers case based insights that highlight: 1) a reactive approach to skills system development for the environmental sector at a systemic level; 2) inadequate differentiation of environmental/sustainable development occupations and of sustainable development practice demands within occupations; 3) a history of dualism between the environment and development; and 4) various broader issues of poor quality education, social marginalisation, organisational inefficiencies and work and learning knowledge acquisition patterns, amongst others. There is also an ongoing need to locate all these systemic developments and challenges within the wider political economy, and political ecologies of the time (discussed further below), and care needs to be taken to avoid conflation of problems with skills system development with problems associated with the prevailing political economy (Allais, 2014), while also appreciating the relationality of these inter-related systems.

Type 2 studies: Related to Type 1 studies, and often embedded in them is a second type of study which involves *development of methodologies and approaches to green skills research that produce analysis of how system elements are related to each other and/or that reveal absences in the system*, as can be seen from a number of the chapters in this book. These become particularly visible when a laminated system perspective is used (cf. Chapters 1, 3, 6, 7, 8, 11 and 13). We have found that a critical realist dialectical orientation foregrounding absence and emergence in skills system research is helpful for showing gaps and absences that require attention to enable a more integrated, joined up system that, for example, links up career information with the occupational classification system, qualifications and provisioning system, emerging demand insights with supply side responsiveness and moving beyond the dominance of a high skill focus in the sector, amongst others.

As can be seen, the research explored across this book is relevant to different 'parts' of the education-training-workplace system. Figure 14.1 draws on Bhaskar's (2010) seven scalar laminated framework (cf. Chapter 13) to consolidate emerging elements of green skills which are named below in the figure as analytically separate but, in reality, are systemically whole consolidating the *interrelated meta elements* emergent across this book. Figure 14.1 consolidates this perspective on green skills research in that it depicts important elements at various levels: the global societal level, the level of post-school education and training, the level

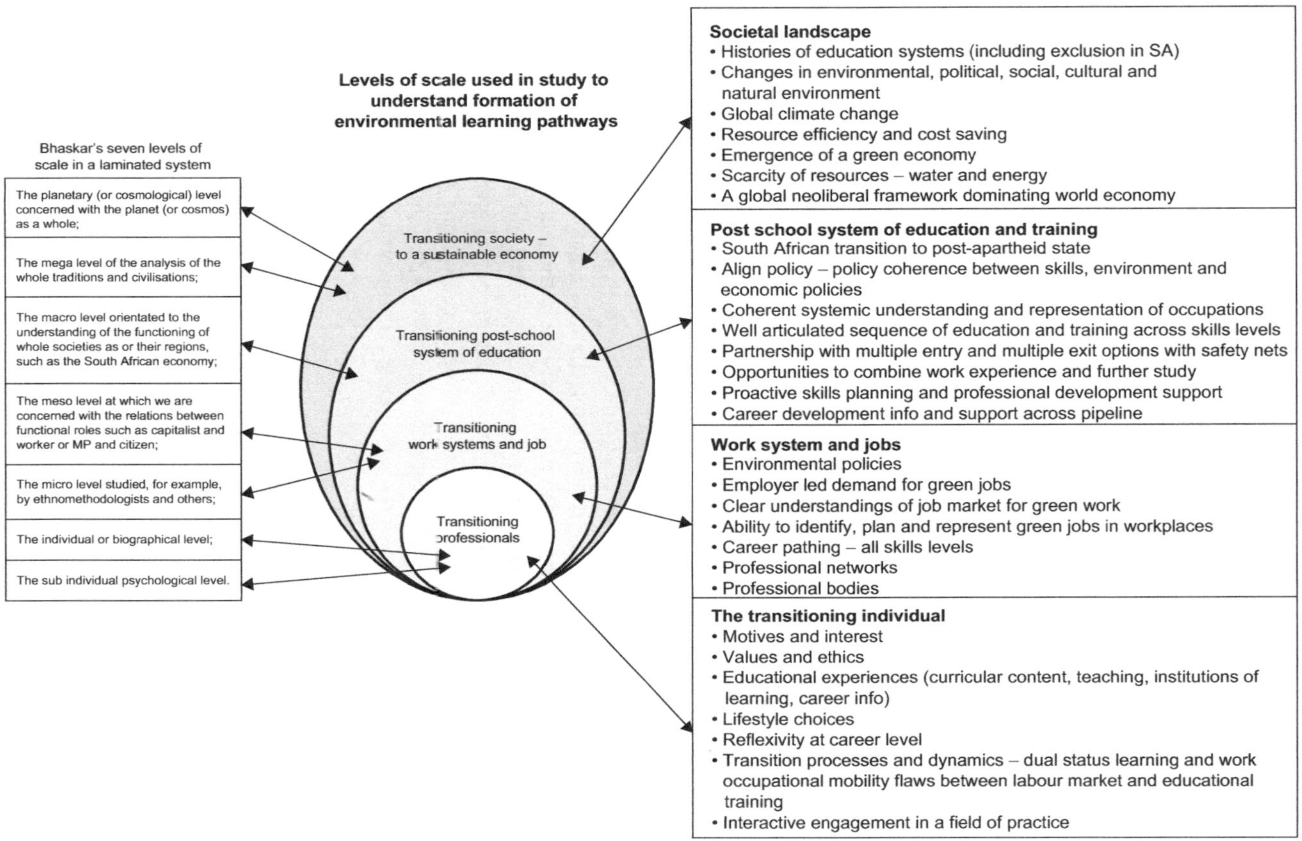

Figure 14.1 A laminated analysis and green skills system elements
Source: Ramsarup (2017, p. 290, drawing on Bhaskar, 2010)

of work systems at employer levels and, lastly, the individual level. The figure highlights that at the level of societal landscape, green skills learning pathways are influenced by the environmental condition itself (cf. Chapter 1), new legislation and compliance demands, resource efficiency and cost saving, emerging opportunities associated with the green economy (cf. Chapter 2) and historical and contemporary political economies of education and environment, as can be seen across the chapters of the book.

Type 3 studies: Building on Type 1 and 2 studies is a third type of study which is relevant to the emerging green skills learning landscape, namely studies that are *more systemically framed* which require *larger scale research programmes that address the need for system wide, proactive skills development planning bridging and bringing together system elements in relational perspective*. Examples of such systemic studies are the 2010 DEA Environmental Sector Skills Planning Study that required investigation of diverse system elements, as well as tools for relational and absence analysis (i.e. the two types of studies outlined above), the South African Qualifications Authority (SAQA) Phases 1 and 2 Sustainable Development Practices Learning Pathways Study conducted over an eight-year period (Lotz-Sisitka et al., 2017) that also required this combination of Type 1 and Type 2 studies; and the green skills value chain demand led studies conducted by the Green Skills Programme over a period of five years, which also required these Type 1 and Type 2 studies (cf. Chapters 2, 3 and 4, www.greenskills.co.za). In all of these cases, we see that conducting Type 1 and Type 2 studies alone is inadequate for producing the level of policy and system development synthesis that is required for substantively advancing the field of green skills, especially if the objective is to make the research more visible in the policy system and landscape. Important in this cluster of more systemic, national and international level studies is the need to include a lens on the emergence of green skills in wider skills system development in national systemic evaluations and skills system research interventions such as research being undertaken to inform the National Skills Development Strategy or research being undertaken to inform the national system of Labour Market Intelligence.

Critical issues emerging for skills system development

The critical issue that emerges from the review of Type 1 studies is that there is *no clear mechanism for sharing the insights from studies such as the ones above into the ongoing skills planning and development system* (for example, these studies have not impacted on the Labour Market Intelligence system of skills research and planning as yet) because green skills are still poorly understood and represented in mainstream skills policy. With the advent of the Sustainable Development Goals (SDGs) as policy driver this may change, and the challenge will then be to synthesise the key insights from these studies to inform this recent policy development.

A critical issue that emerges from the review of Type 2 studies is that *there is a need to develop stronger integrative frameworks and tools for skills system development as recommended in the Department of Higher Education and Training (DHET) Post-Schooling White Paper of 2012* (Sauls, 2018). This requires green skills research that is held together by ongoing, field reflexive processes of synthesis and co-ordination which is what the DEA (2010) Environmental Sector Skills Plan and the Development Bank of South Africa (DBSA) funded Green Skills Research and Development Programme (captured in this book) set out to do. However, the policy environment is ever-changing (cf. Chapter 2), and it is clear, in synthesising the work across this book, that there is need for a new generation of green skills research that builds on the methodological insights shared across this book in ways that can offer a new level of national synthesis within the new policy environment of the SDGs, and the revised National Development Plan which is currently in process in South Africa. Overall, there is need to better align the green skills trajectory in South Africa with the (revised) National Development Plan's intentions and the SDGs. Perhaps there is an immanent need for a 'National Skills Plan for Sustainable Development'? While some gains have been made towards a more proactive system of green skills planning and provisioning, these gains still remain inadequate for the scope of the challenges of the just transitions necessary for sustainable futures.

A critical issue to be addressed out of the third type of study is *the problem of translation of research into skills planning and practice interventions*. We have found over the years that cross sectoral concerns such as environment and sustainability don't have adequate national synthesis mechanisms in the skills system, unlike other transformational imperatives (e.g. youth unemployment or HIV/AIDS). This is because the sustainability cross sectoral issues don't have a specific issue focus that is easy to isolate and bring into focus. This also makes it difficult to generate data on the inclusion of sustainable development into the skills system, and it is clear that there is need to establish a better mechanism for dealing with cross sectoral issues such as the SDGs so that these can be included in regular monitoring and evaluation systems and in skills planning logics and processes. As was shown in the South African Qualifications Authority (SAQA) research programme (Lotz-Sisitka et al., 2017), this will necessitate transversal engagement as sustainability practices are located across schooling, higher education, occupationally directed training and wider social learning.

All of the *types of studies* outlined above, along with the methodological innovations that have accompanied them, and the critical issues surfaced from our analysis, illuminate what is involved in integrating a new knowledge field into a national system of skills development. This gives rise to our policy recommendations, outlined below, all of which are critical for sustainable development in South Africa. Ultimately green skills research gives meaning to the clause in the South African Constitution (1996), which guarantees the right to a healthy environment and sustainable development of South Africa's resources for current and future generations (cf. Chapter 2). In this sense, green skills research is not a 'nice to have', but is integral to protecting the rights and-well being of society, now and in future.

Policy implications

Adopt a more systematic approach to aligning green skills with SDGs and other policy imperatives

As can be seen across the chapters in this book, there is need to develop a more systemic approach to green skills research and planning in South Africa, at the highest level and across all sectors. Through this, the SDGs need to come into focus in the national landscape of skills development, building on what has been learned from environment and sustainability oriented green skills research and planning since the early 1990s (cf. Chapter 1). The ambition expressed in the South African National Development Plan, the African Union 2063 Agenda and the UN SDGs underscores the urgency for societies to transform towards a more liveable, just and ecologically sustainable future. In such a context, transformation is not a buzz word, nor is it ambiguous. It is clearly defining the era and way in which we live. Neglecting to develop adequate mechanisms for advancing green and sustainable development related skills development would neglect an important dimension of transformation of South African society, and society more broadly.

As argued in Chapter 1, green skills research, if conceptualised within a just transitions and transformative framework in South Africa, aligns with the SDGs, and the National Development Plan's (NDP's) aims to eliminate poverty and reduce inequality by 2030, hence we have emphasised not only the high skills pathway which currently dominates green skills development in South Africa, but also new models for skills planning that can maximise the investment currently being made into skills planning in entry level occupation and programmes such as the Expanded Public Works Programme (cf. Chapter 8), as well as more effective skills planning to develop relevant occupational families and tools for advancing a wider range of occupations for greening work at all levels of the system (cf. Chapters 3 and 12), and to undertake research that can also identify latent demand for green skills (cf. Chapters 3, 4 and 5), as this can potentially also unlock new work streams and occupations in a rapidly emerging field. Here re-skilling of the existing workforce into cleaner and more decent jobs is also an important dimension of the argument being made across this book. We make this argument with a commitment to transformative learning and curriculum transformation (cf. Chapters 10 and 11) as neglecting the transformation needed in education and training institutions would be a fatal ommission in conceptualising transformation in the skills system. As such, in providing insights into how to improve the quality of skills training and planning in the country, we also recognise the importance of the NDP's view that poor quality education is a 'structural impediment' to the country's development. Our book argues that this needs to be done with the full force of the national system of skills development, and that this requires more joined up thinking, systemic planning and stronger, more proactive mechanisms for articulating and designing green skills and sustainable development learning pathways for all.

In focusing in on the SDGs in an integrative, critical and transdisciplinary way via green skills learning pathways and skills system development, we recommend not only skills development for specific SDGs (for example, SDG 13 on climate action, or SDG 6 on water and sanitation), but also a focus on absences and weaknesses in the SDGs (for example, their targets for community participation are inadequately constituted; and there are underlying contradictions across the SDGs). We also propose a strong and critical emancipatory focus on the cross cutting SDGs that build social competence and green skills for driving the other SDGs. In SDG 4, Target 4.7 specifically requires Education for Sustainable Development to be integrated into the lifelong learning system (schools, universities, colleges and informal and non-formal learning programmes). It requires that all countries "ensure that all learners acquire the knowledge and skills needed to promote sustainable development [by 2030]". Here United Nations Education, Science and Cultural Organisation (UNESCO) (2016) underscores that education is both a goal in itself and a means of attaining all the other SDGs, which positions education, training and skills development as a key enabler or driver of sustainable development.

To link skills development system outcomes reflexively with the development trajectories outlined above, there is also need to include a complementary focus on monitoring and evaluation in which green skills trajectories and skills for sustainable development and just transitions are integrated into national systems of monitoring and evaluation. Currently, there is little guidance on how to construct relevant, situated indicators for monitoring of green skills for sustainable development in co-engaged ways, pointing to another dimension of policy engagement and development out of the history of green skills research. Overall, the perspectives above all add impetus to our recommendation for an immanent national synthesis of 'Skills for Sustainable Development' that can inform policy and practice.

Measures for shorter and longer term systemic engagement and impact

There are a number of measures that are required to enable longer term systemic engagement and impact relevant to advancing green skills research and planning. These include:

- **Adopt systemic approach:** As outlined above, and despite the extensive and complex green/sustainable development mandate in policy (cf. Chapter 2), to date no adequate co-ordination mechanism has been established for the national planning and development of green skills in South Africa (Rosenberg et al., 2016), and at post-schooling education training and (PSET)level, to date there is no coherent analysis and coordination of the skills needs linked to sustainability. Current inadequacies in green skills learning pathway development and the inability of the education system to meet green skills needs have not been coherently discussed within education and training system policy forums in South Africa. This has effects in terms of demand and supply analysis and provisioning, and is leading to substantive mismatches and a general situation of inadequate coherence as has been illustrated across this book.

- Policy, programmes and training exist in pockets across the country but are not co-ordinated to work together. At all levels of government, non-governmental organisations (NGOs), unions and education centres are all working on their own green skill programmes leading to a proliferation of green jobs analyses and human capital development strategies, many of which are overlapping and/or 'double counting'. These are yet to be coordinated to work together in a broader context within the wider economic development strategy and policy framework. Thus, we have a paradox of support for green skills seemingly appearing everywhere, but actually being partly invisible in the national landscape.
- There are a number of different structures that can be engaged, for shorter term gains, for example, training of Sector Education and Training researchers in green skills research planning, and use of the tools, for example, supporting the nationally developed sourcebook focusing on green skills planning in and across a variety of sectors (Ramsarup et al., 2017). For longer term gains, there is need to align with national policy for just transitions and implementing the SDGs, along with dealing with backlogs in implementing national environment and sustainable development policy (for example, for waste management, water security and more). We have argued across this book that *just transitions for sustainability need proactive green skills planning*, and there is need to collect the cumulative insights gathered at niche levels in the system into more systemically synthesised perspectives that can be amplified for wider impact (i.e. by combining the three types of research outlined above).
- **Adopt a policy-in-practice approach to develop skills systems:** As indicated one of the challenges of green skills research is that much of what needs to be done is not always known in advance, as the environment and sustainability policy landscape has been rapidly emerging alongside emerging experience and knowledge of green skills research (cf. Chapter 2). This requires that we adopt *a policy-in-practice approach* to environmental and sustainable development research and learning. This allows possibilities for research to influence policy and practice, and provides a good argument for generative, co-engaged research for skills system building (cf. Chapters 11 and 13). For example, research into green occupations undertaken by Ramsarup (2017) raised the problem of almost no focus on green occupations in the Organising Framework for Occupations (OFO). Engagement with the policy system

on this problem during the research process led to integrating green occupations from the American Occupational Information Network (ONET) system into the South African Organising Framework for Occupations in 2015. However, while this was an advancement on what was there before, the American categories of occupations did not always match the South African contextual realities. This required further research into occupational families and how they are constituted (cf. Chapters 5 and 12) that can now be used by all sectors to clarify occupational descriptions for green work, the greening of work and green jobs (Ramsarup & Ward, 2018).

Implications for work and learning

Political economy meets political ecology

Not only does green skills research implications for skills research and policy, it also introduces new perspectives into critical research that has traditionally underlaboured transformative skills research, to date mainly framed from a political economy perspective as labour rights, labour relations and well-being became a strong focus for critical research, shaped also by Marxian thinking over the past 200 years. This body of research has shaped discourses on labour, work and vocational education and training worldwide (for example, Brown et al., 2001; Brown et al., 2011; Lauder et al., 2012; Livingstone & Guile, 2012). With increased attention being given to the immanence of ecological collapse, and the implications of this for the majority of the world's people, better understandings of the critical tipping points that are immanently associated with global warming and the massive risk that this presents to societies all over the world, the political economy is giving attention to 'green issues' (Moore, 2015), and the limits of capitalism are being articulated not only in relation to exploitation of labour and labour relations, but also exploitation and appropriation of life support systems and planetary commons (e.g. water, air, land, etc.).

From this perspective, the traditional contradiction of use value and exchange value is being deepened with an understanding that we are facing a related, but immanently more complex central contradiction, that of *commodification value vs co-existence value*. For example, we can see that if water becomes over-commodified, it reduces the capacity for co-existence of all, where there ought to be some water for all, including the non-human species that rely on water for their survival. There is also a large body of research emerging around the origins, limitations and impacts of 'fossil capital'. For example, Malm (2016) articulates how, in the early years of the Industrial Revolution, manufacturers turned away from traditional sources of power (water mills) towards coal power, not because it was cheaper, but because it afforded more control of subordinate labour. These political ecology histories are showing how, especially the control of energy, but also water, minerals and land (for example, via colonial resource extractivism) were used to concentrate resources in ways that benefitted the powerful and that fuelled the colonial project (Martinez-Alier, 2002). This led to neo-liberal concentrations of production at the most profitable sites, with recent histories showing that this included transferring pollution 'off shore' to developing countries with lower environmental standards, and with subordinate labour with little power to set the agenda for clean and decent work (ibid.).

The emergence of discourse around 'just transitions' (Swilling & Anneke, 2012, cf. Chapter 1) is therefore not only a discourse about more sustainable options for future generations, it is also deeply political and addresses the heart of contemporary political economy debates. One can consider a range of examples here, the stalling of shifts towards renewable energy being an interesting case, which provides options for the devolution of power that can also work towards overthrowing the current economic order and its centralised historical power structures – over both resources and labour. This context positions green skills research at the interface of the emerging synergies between political economy and political ecology.

In the Prologue, we noted that this is a significant framing for contemporary green skills research, and throughout the book we have shown how underlabouring studies with critical realism deepens critical analysis and understanding of skills system studies and the way in which the skills system is including and/or excluding people and other forms of life. This broadens notions of work to include market-sector work, household sector work, state-sector work and work for the commons (Raworth, 2017) and ushers in a concept of economy that is not just profit oriented, but oriented towards re-generation of life, social cohesion, cultures and societies.

The nature of occupations and work

As indicated above and in Chapter 1, the wider societal shifts towards taking sustainability concerns more seriously within the wider zeitgeist of humanity's trajectory has given rise to new politics, policies and social movements for transformations towards sustainability, with implications for the nature of occupations and work, as well as wider social learning. Transitioning to a greener more sustainable and just society has focused primarily on shifting technologies and policy, especially in the branch of transitions theory that emphasises socio-technical transitions (Geels & Schot, 2007; Geels & Kemp, 2012). In the just transitions discourse (Swilling & Anneke, 2012), work and its role in society needs to be re-examined, and a stronger focus on workers and how they are included or excluded needs to be brought into focus (cf. Chapters 2 and 8). Attention to how the socio-technical transitions (for example, towards renewable energy) will change the nature of work is rare in the transitions field, and an analytical construct that provides insights into how the nature of work itself is/will change is even rarer.

Another implication is how work is conceptualised, involving at minimum the integrated nature of productive and reproductive activity. Occupational discourse has tended to focus more on productive work, neglecting reproductive aspects of work – or the social side of work. We have argued throughout this book that green skills introduce a *regenerative* dimension of work (cf. Raworth, 2017). Shalem and Allais (2018) argued that work has normative, epistemic and social dimensions. We have also argued that research on work should further address the ontological dimension of work, seen most obviously in the presence of environmental relations, but also in ontological dimensions of institutions and subjects themselves, as this also allows for considering the *regenerative dimensions* of work in society.

In the discussion below we draw on the Shalem and Allais (2018) framework and explore some of its implications from an environmental or green skills vantage point. Green skills research points to green work having normative, epistemic, social and ontological dimensions that define the nature of green work and learning (see Figure 14.2).

In brief, the *normative dimensions* of green work are related to the way in which work ethics are being transformed towards a more regenerative paradigm for people and planet relations. Bauman (2004) explained how the modernist industrial system and mode of work transformed work ethics from a community-oriented work ethic to a decontextualised work ethic required for factory efficiency, which was only relevant to the workplace itself. Sustainability re-frames work ethics towards wider contextual and planetary questions, including immediate public

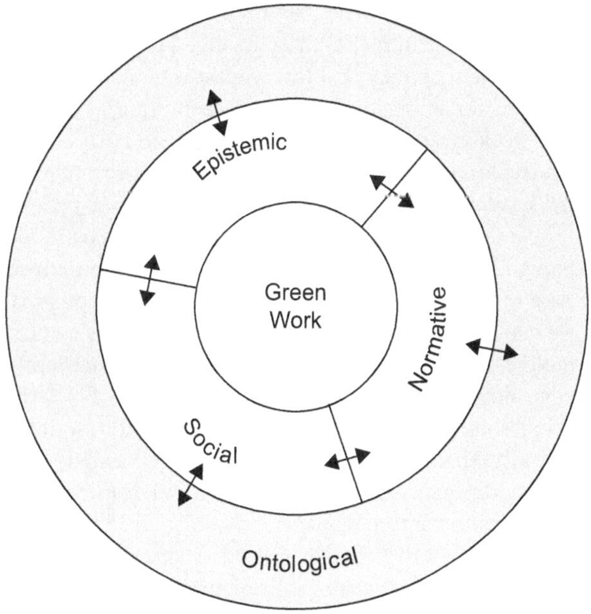

Figure 14.2 Conceptual framing of green work with normative, epistemic, social and ontological dynamics

and environmental health concerns (for example, the implications of pollution for workers' health and others' lives) and questions around current and future generations and their well-being. It is often said that in contexts of poverty, people are not concerned with such ethics, but this is a gross over-simplification of human capabilities and experience and says more about the context in which such interpretations are being made, rather than the choices of people given viable alternatives and options from which to choose. The normative dynamic of sustainability also brings new practices into being, which find relevance in the workplace such as energy efficiency, ecological restoration and so on, which in turn also informs our focus, and the differentiation developed in this book, on the greening of work, green work and green jobs (cf. Chapter 12). Normative dynamics also drive the emergence of new work in sectors, as shown in the efforts to develop new tools for demand analysis in the first part of this book. As shown in Chapters 6–12 in this book, this has implications for how people are experiencing work, transitions into work and preparation for work, and it also shapes provision of education and training and possibilities for transformative learning. Learning processes and programmes are also shifting with a strong call for ethics-led, co-engaged, transformative and even transgressive forms of learning (cf. Chapters 10 and 11).

From an *epistemic perspective*, green work involves engaging with new knowledge, and that which is not yet known, often still only articulated in emerging scientific texts (for example, climate change projections), which are also uncertain in their projections. This is because much environmental science knowledge is predictive. There is therefore need to articulate the implications of this knowledge for emerging practice (for example, climate change projections influence planning of the renewable energy policy and interventions, which in turn have new knowledge implications for skills development and training). For example, this could influence the work of an electrician, who needs to both understand and adapt to changes in the knowledge environment as they shift from installing fossil fuel driven energy systems to solar energy systems. This can happen at any time (for example, mid-career) and it often involves unlearning existing knowledge and learning completely new principles that underpin the knowledge itself, as was shown in the case of the vocational college lecturers in Chapters 10 and 11. In this example, we can see that environmental science knowledge produces demand for new technology and practice, which in turn produces new epistemic contexts and knowledge flow challenges (cf. Chapters 10 and 11). Another challenge currently facing South Africa, i.e. how to adapt to conditions of severe water scarcity (i.e. all South Africa's water has already been allocated) is producing a new knowledge environment in the water sector which introduces new concepts such as ecological infrastructure management and more. These concepts and associated skills demands need to become embedded across the entire water value chain, which has implications for green skills planning, especially if it is to be more pro-actively constituted than in the past.

From a *social* point of view, much of the transformation required towards just transitions requires a joined up, wide and substantive social and cultural change process, including in work and learning cultures. As articulated across the chapters in the book, there is need for a stronger emphasis on the social dynamics of environmental change, including an emphasis on collaborative learning and agency for change (cf. Chapter 11), development of workplace cultures that value the environment/ecological integrity (not only profit and extraction) and more. Just transitions also bring historical analysis and perspective to the fore; in order to better understand why new sustainability practices are needed in society, there is need to also understand problems of past development trajectories such as the impacts of fossil fuel on the climate system, and associated implications for current and future generations, i.e. it brings the past-present-future into relationships in the social dynamics of economy and work.

Across the book, we have also argued that green skills and greening of work is also *ontological*, in that there is need to make more visible the ontological dynamics of green work (cf. Chapter 13). For example, the ontological implications of chemical pollution for water security can help to shape green work and green jobs that are oriented towards pollution control or monitoring. It is not only the physical or ecological ontological dynamics that are important to green work and learning, there is also need to surface the real via examinations of the embodied and constructed nature of work and the multiple factors that contribute and fail to contribute to shaping work and who and what is determining the nature of work. It asks that we do not leave work to the neo-liberal economy, and make ontological arguments for including the currently and increasingly marginalised interests of society and the planet, including the well-being of other life forms. Importantly the surfacing of the latent, i.e. types of work that are not (yet) visible, and what is absent are important for the emergence of green skills; these are as important as identifying the deep structures that produce or hold the absences in place (i.e. an absence of ecological economics curricula in higher education institutions).

Conclusion

In this chapter, we have tried to synthesise some of the implications of our research for systems, policy and for understanding work and learning. We have shown the different types of studies that are needed to drive wider systemic integration of green skills research into the national system of skills development and policy alignment. We have argued for a new level of synthesis in the light of the recent promulgation of the SDGs and the immanent review of the NDP, contextualised in the longer term history of environmental policy and skills system policy and praxis to ensure stronger alignment between environmental policy and education system policy and development (cf. Chapter 2). We have identified some critical issues that need addressing. From a wider point of view, we have considered the way in which political ecology is offering new perspectives to political economy theory, which for a long period has driven critical education and skills research trajectory. With this in mind, we have also briefly explored the nature of green work and learning from a normative, social, epistemic and ontological point of view. Overall, the chapter seeks to bring the diverse contributions together into a pathway for new policy development research, as well as new forms of critical skills research and ongoing engagement with education and training system development.

References

Allais, S. (2014). *Selling out education. national qualifications frameworks and the neglect of knowledge*. New York: Springer.

Bauman, Z. (2004). *Work, consumerism and the new poor*. Milton Keynes: Open University Press.

Bell, C. (2019). *Understanding the learning that occurs through up-skilling opportunities and practices in the marine sector of South Africa*. Unpublished PhD research thesis, Rhodes University, Grahamstown, South Africa.

Bhaskar, R. (2010). Contexts of interdisciplinarity and climate change. In R. Bhaskar, C. Frank, K.G. Hoyer, P. Naess, & J. Parker (Eds.), *Interdisciplinarity and climate change: Transforming knowledge and practice for our global future*. London and New York: Routledge.

Brown, P., Green, A., & Lauder, H. (2001). *High skills: Globalization, competitiveness, and skill formation*. Oxford, and New York: Oxford University Press.

Brown, P., Lauder, H., Ashton, D.N. (2011). *The global auction: The broken promises of education, jobs and incomes*. New York: Oxford University Press

Department of Environmental Affairs (DEA). (2010). *Environmental Sector Skills Plan for South Africa*. Pretoria: DEA.

Dotwana, A.S. (2015). *The factors that constrain or enable black female Botany and Zoology honours students in making the transition to Masters studies in Eastern Cape universities*. Unpublished Master's thesis, Rhodes University, Grahamstown, South Africa.

Geels, F.W., & Schot, J. (2007). Typology of sociotechnical transition pathways. *Research Policy*, 36, 399–417.

Geels, F.W., & Kemp, R. (2012). The multi-level perspective as a new perspective for studying socio-technical transitions. Automobility in transition. In F.W. Geels, R. Kemp, G. Dudley, & G. Lyons (Eds.). *A socio-technical analysis of sustainable transport*, pp. 49–79. London: Routledge.

Lauder, H., Lauder, P., Young, M., Daniels, H., Balarin, M., Lowe, J. (2012). *Educating for the knowledge economy? Critical perspectives*. London: Routledge.

Livingstone, D.W., Guile, D. (2012). *The knowledge economy and lifelong learning: A critical reader*. Rotterdam and Boston: Sense Publishers.

Lotz-Sisitka, H., Ramsarup, P., Bolton, H., & Reddy, J. (2017). Introduction: Researching sustainable development learning pathways towards progression in learning and work. *SAQA Bulletin: Learning pathways, the National Qualifications Framework (NQF), and lifelong learning in South Africa*. *17*(1), 10–30.

Malm, A. (2016). *Fossil capital: The rise of steam power and the roots of global warming*. London: Verso.

Martinez-Alier, J. (2002). *The environmentalism of the poor. A study of the ecological conflicts and valuation*. Cheltenham: Edward Elgar.

Moore, J. (2015). *Capitalism in the web of life: Ecology and the accumulation of capital*. London: Verso.

Olvitt, L. (2012). *Deciding and doing what's right for people and planet: An investigation of the ethics-oriented learning of novice environmental educators*. Unpublished PhD thesis, Rhodes University, Grahamstown, South Africa.

Pesanayi, T. (2019). *Boundary crossing expansive learning across agricultural learning systems and networks in southern Africa*. Unpublished PhD thesis, Rhodes University, Grahamstown, South Africa.

Ramsarup, P. (2017). *A critical realist dialectical understanding of learning pathways associated with two scarce skill environmental occupations within a transitioning systems frame*. Unpublished PhD thesis, Rhodes University, Grahamstown, South Africa.

Ramsarup, P., Ward, M., Rosenberg, E., Jenkin, N., & Lotz-Sisitka, H. (2017). *Enabling green skills: Pathways to sustainable development: A source book to support planning for green economies 2017*. doi: 10.21504/10962/62694

Raworth, K. (2017). *Doughnut economics: Seven ways to think like a 21st-century economist*. London: Random House Business Books.

Republic of South Africa. (1996). *Constitution of the Republic of South Africa*. Pretoria: Government Printers.

Rosenberg, E., Ramsarup, P., Lotz-Sisitka, H., & Gumede, M. (2016). Building capacity for green, just and sustainable futures – a new knowledge field requiring transformative research methodology. *Journal of Education*, 65, 95–122.

Sauls, G. (2018). *The National Skills Fund and green skills: Towards a generative mechanism approach*. Unpublished PhD thesis, Rhodes University, Grahamstown, South Africa.

Shalem, Y., & Allais. S. (2018). *Linking knowledge, education and preparation for work: Exploring occupations*. Working Paper November 2018. Centre for Researching Education and Labour, University of Witwatersrand.

Songqwaru, Z. (2019). *A theory-based approach to evaluating a continuing professional development programme aimed at strengthening environment and sustainability education*. Unpublished PhD thesis, Rhodes University, Grahamstown, South Africa.

Swilling, M., & Annecke, E. (2012). Just transitions: Explorations of sustainability in an unfair world. Cape Town: University of Cape Town Press.

United Nations Educational, Scientific and Cultural Organisation (UNESCO). (2016). Global education monitoring report. Education for people and planet. Creating sustainable futures for all. Paris: UNESCO. Retrieved from https://en.unesco.org/gem-report/report/2016/education-people-and-planet-creatingsustainable-futures-all

Ward, M., McClean, D., Kraak, A., Jenkin, N., & Mushangai, D. (2017). *The eco-furniture programme. An evaluative review*. Pretoria: Department of Environmental Affairs.

Index